J. ALLEN HYNEK: THE MAN WHO COULDN'T LIE

Allen Schery

Brooklyn Bridge Books

Brooklyn Bridge Books

Contents

Prologue — The Press Conference

Prologue — The Press Conference

The room is too loud and too warm and J. Allen Hynek is standing at a podium he did not want to stand at, in a city he did not want to visit, about to say something he does not quite believe.

Outside the Detroit Press Club on March 25, 1966, the last of a Michigan winter is retreating into gray slush. Inside, several dozen reporters have packed themselves into a conference room with inadequate ventilation and an abundance of skepticism. They have notepads. They have cameras. They have been waiting for days, ever since the sightings started, ever since the phones at county sheriff's offices began ringing in the middle of the night and didn't stop. They are not here to be patient.

Hynek adjusts his glasses. He is fifty-six years old, trim, professorial, with the careful bearing of a man who has spent three decades in university lecture halls. He has a beard that makes him look vaguely European, a pipe he carries more than he smokes, and an appointment at Northwestern University where he chairs the astronomy department and is respected by people whose respect he values. He is not, by any measure he would have chosen, a man who belongs in this room.

And yet here he is. Here he has been, more or less, for eighteen years.

The Air Force calls it Project Blue Book. His official title is scientific consultant, a designation that sounds more prestigious than it is. What it means in practice is that when something embarrassing happens — when the phones ring and the newspapers gather and the public begins to ask uncomfortable questions — Hynek gets on a plane. He looks at the evidence. He talks to the witnesses. He finds an explanation that allows everyone to go back to normal. He has done this many times. He is good at it.

He is about to discover the limits of how good.

It began, as these things often do in Michigan in March, in the dark. On the evening of March 20th, a young woman named Donna Marcum and her family reported seeing something low over the swampy ground near their farmhouse outside Dexter. Two days later, eighty-seven students and a dozen civil defense workers at Hillsdale College watched something hover silently over a field for four hours. In between, dozens of other witnesses — farmers, motorists, a deputy sheriff who had been skeptical of such things his entire professional life — reported variations on the same experience. Something bright. Something silent. Something that moved in ways that aircraft do not move and then was gone.

By the time Hynek arrived the case had developed a velocity of its own. Gerald Ford, then a congressman from Michigan, had called for a congressional investigation. The wire services had picked it up. The Air Force, which had an institutional allergy to congressional investigations, wanted a resolution. They wanted it quickly, and they wanted it to be the kind of resolution that ended conversations rather than started them.

Hynek spent two days in the field. He talked to witnesses. He walked the marshy ground near the Marcum farm in his city shoes. He looked at what there was to look at, which was not much, because whatever had been there was not there anymore. The witnesses were not the kind of people he had learned to dismiss — they were specific, consistent, unfrightened by the attention, more puzzled than dramatic in their accounts. The deputy sheriff described what he had seen with the clipped precision of a man filing a report.

None of this was necessarily surprising. In eighteen years of Blue Book consultancy Hynek had talked to many credible witnesses about things he couldn't explain. He had learned to sit with that discomfort, to write his notes and file them and move on. But the Air Force needed something to tell the press and the press needed something to tell the public and the public needed something to tell itself before the next story came along.

He thought about it on the drive back from Dexter. He thought about it in his hotel room. He thought about a phenomenon he had read about — marsh gas, methane released by decaying organic matter in wetlands, sometimes luminous, sometimes mistaken at night for something more exotic. It was not a new theory. It was not a particularly strong theory for what the witnesses had described. But it was a theory, and it was attached to a phenomenon that was real, and it was the kind of thing a scientist could say to a room full of reporters without being immediately laughed out of it.

He wrote it into his notes. He told himself it was a possibility. He did not tell himself it was the answer, because he did not believe it was the answer. But possibility and answer are different things, and in eighteen years he had become practiced at the distance between them.

The cameras are on now. The questions are already beginning before he finishes his opening remarks — shouted from the back of the room, overlapping, not particularly interested in his qualifications or his methodology or the careful epistemological hedges he has prepared. They want the explanation. They want it simple. They want it now.

He gives them marsh gas.

He explains the phenomenon carefully, technically, the way he would explain it to a graduate seminar. The organic decay, the methane, the luminosity, the visual distortion possible at night in low-lying terrain. He watches the room as he speaks. He watches the reporters writing it down, already simplifying, already reducing the technical explanation to the two words that will appear in every headline by morning. Swamp gas. Somewhere in the middle of his own explanation he becomes aware of a particular quality of silence. Not the silence of a room that is listening. The silence of a room that has already decided. Several reporters have stopped writing. A man in the third row has the expression of someone watching a performance he expected to be worse than it is and finding it is exactly that bad. At the back of the room someone laughs — not loudly, not cruelly, but with the involuntary brevity of a man who cannot help himself.

Hynek continues. He is a professional. He finishes what he started.

Afterward, driving to the airport in a car arranged by the Air Force, he looks out the window at the flat gray Michigan landscape and does not think about marsh gas. He thinks about the deputy sheriff's face when he described what he had seen. The specific, unhurried way the man had spoken, with no investment in being believed, just a policeman describing an incident the way policemen describe incidents. He thinks about the students at Hillsdale and the four hours they had stood in a field watching something that by every account they had tried to explain away themselves before they gave up.

He thinks about the word possibility and the word answer and the space between them where he has been living for eighteen years.

He is fifty-six years old. He is chairman of the astronomy department at Northwestern University. He has a pipe in his coat pocket and a flight home to Chicago and a lecture to prepare for Monday morning. He has spent nearly two decades explaining things away for an institution that needed them explained away, and he is good at it, and it has cost him something he hasn't fully named yet.

The car moves through the gray afternoon. Hynek watches Michigan go past. Something, he thinks, has to change.

He does not know yet what it will cost him to change it. He does not know that the two words already running across the wire services will follow him for the rest of his life, appearing in introductions at conferences and punchlines in newspaper columns and the mouths of people who want to dismiss him. He does not know that the discomfort he is feeling in the back of this Air Force car is the first clear signal of a transformation that will redefine his career, his reputation, and his understanding of what science is actually for.

He knows only that he stood at a podium and said something he did not believe, and that he has done versions of this for eighteen years, and that the deputy sheriff's face is still in front of him.

He is, at his core, a careful man. A precise man. A man who was trained to follow the evidence and say what the evidence says and nothing more and nothing less.

He has not been doing that.

The airport appears at the edge of the flat horizon. The car slows. Hynek picks up his bag from the seat beside him, checks that his pipe is in his pocket, and steps out into the cold. Behind him, in a marshy field outside Dexter, Michigan, whatever was there is gone.

What follows is the story of what it cost a scientist to tell the truth — and what the truth turned out to be.

Chapter One

An Ordinary Life in an Ordinary City

Josef Allen Hynek is born on May 1, 1910, in Chicago, Illinois, which is not yet the city it will become but is already becoming it. The stockyards are running. The elevated trains are running. The lake sits at the eastern edge of everything, cold and enormous and indifferent to human ambition, which in Chicago runs at a considerable volume. It is a city of arrivals — people from everywhere coming to become something, bringing with them languages and foods and memories of places they have left behind and are trying not to miss too obviously. The Hyneks are among them. His parents are Czech, part of the wave of central European immigration that settled the neighborhoods on Chicago's north and west sides and built communities that were simultaneously loyal to the new world and quietly homesick for the old one. They are not wealthy. They are not poor. They are the particular kind of solid, striving, education-valuing family that produces scientists and lawyers and engineers and considers this a reasonable outcome for a life.

The neighborhood is Pilsen, on the lower west side, which in 1910 is densely Czech and will remain so for decades. The streets have a specific texture — brick two-flats with small front stoops, the smell of bakeries in the morning, the sound of a language that is not English coming through open windows in summer. It is an enclosed world in the way that immigrant neighborhoods are enclosed, not by hostility exactly but by sufficiency — everything needed is here, the church, the market, the school, the social club, the particular comfort of being understood without effort. Young Josef Allen grows up inside this enclosure and it shapes him in ways he will not fully recognize until much later, if at all. The values are clear and consistent: work carefully, think precisely, do not embarrass the family, earn the respect of people whose respect means something.

He is a quiet child by most accounts, more interested in how things work than in the social negotiations of childhood. [*Verify: family accounts, early biographical sources.*] Chicago in the nineteen-tens is a city full of things to be curious about — the mechanical, the electrical, the astronomical. The city has a particular relationship with the sky, perhaps because the grid of its streets is so relentlessly horizontal that the vertical becomes a kind of relief. The Adler Planetarium does not exist yet — it will open in 1930 — but the University of Chicago's Yerkes Observatory sits ninety miles north in Williams Bay, Wisconsin, and its reputation reaches the city's schools and libraries and the minds of boys who are paying attention. Hynek is paying attention.

What draws him to astronomy specifically is difficult to reconstruct with certainty, but the broad shape of it is recognizable. It is the combination of mathematics and mystery that captures a certain kind of precise, wondering mind — the fact that the universe runs on equations you can actually learn, that the distances and temperatures and compositions of objects unimaginably far away can be calculated from first principles by someone sitting at a desk with the right tools and enough patience. There is something deeply satisfying about that to a boy who likes things to resolve cleanly. The universe is vast and strange but it is not arbitrary. It follows rules. Learn the rules and the vastness becomes, if not smaller, at least legible.

He attends the University of Chicago, which in the nineteen-twenties is one of the finest research universities in the country and knows it. The atmosphere is rigorous, empirical, unsentimental about received ideas — the kind of institution that produces scientists who trust data over intuition and evidence over authority. Hynek absorbs this without resistance because it matches something already in him. He is not a rebel. He is not temperamentally drawn to controversy or heterodoxy. He wants to understand things correctly, which means understanding them the way the evidence says they are, which means following the methods that have proven most reliable for getting at what is actually true. This is not a complicated philosophy. It is the philosophy of a careful man, and Hynek is, from the beginning, a careful man.

His field is astrophysics, and within astrophysics his particular interest settles on variable stars — stars whose brightness changes over time, sometimes regularly, sometimes erratically, for reasons that in the nineteen-twenties are still being worked out. It is exacting, methodical work. You observe, you measure, you record, you compare against previous measurements, you look for patterns, you form hypotheses, you test them against new observations. The timescales are long and the increments are small and the satisfaction comes from accumulation rather than revelation. It suits him entirely.

He earns his doctorate from the University of Chicago in 1935, his dissertation bearing the careful, specific title of a man who has learned to make claims only as large as his evidence will support. He is twenty-five years old, newly credentialed, entering a profession during the Depression when academic positions are scarce and competition is considerable. He is good enough to find a place — Ohio State University, in Columbus, where he joins the faculty and begins the slow, steady work of building a career. Publish papers. Train students. Apply for grants. Attend conferences. Establish relationships with the people who matter in the field. It is not glamorous work. It is the work of someone who has decided what kind of life he wants and is building it one careful year at a time.

Columbus is not Chicago. It is flatter, quieter, less electric with ambition, a midwestern university town with a large football stadium and a campus that spreads generously across land that costs less than it would anywhere on the coasts. Hynek settles in. He is not the kind of man who chafes at a place because it is not somewhere else. He has his work and his instruments and his students and the particular pleasure of a mind that is consistently engaged. He is, in the late nineteen-thirties, a young astronomer on a reasonable trajectory toward a respectable career. He will publish papers of moderate influence. He will train students who will go on to their own respectable careers. He will retire eventually with a named lecture hall or a departmental award and a life that added something small but real to the sum of human knowledge. This is not a diminished ambition. It is the ambition of someone who understands what science actually is — not the dramatic revelations of popular imagination but the patient, incremental, collaborative accumulation of carefully verified fact.

Nothing about any of this predicts what is coming.

The war arrives in 1941 and reorganizes everything, as wars do. Hynek ends up at Johns Hopkins Applied Physics Laboratory in Maryland, working on the proximity fuze — a device designed to detonate artillery shells not on impact but when they pass near enough to a target, which requires solving problems of miniaturized electronics and reliable triggering mechanisms under conditions of extreme stress It is applied science of a practical and urgent kind, far removed from variable stars and long observational timescales, and it demonstrates something about Hynek that his academic career has not required him to show — that he can move between the theoretical and the practical without losing his footing in either. He is useful. He is adaptable. He is the kind of scientist that institutions reach for when they need a problem solved by someone who will not be distracted by the elegance of the solution.

He returns to Ohio State after the war and continues. More papers, more students, more of the careful incremental work that constitutes a scientific life. He modernizes the instrumentation at the university's observatory, improves the photometric methods his students use to measure stellar brightness, contributes to the slow communal project of understanding why certain stars vary and others do not.He is in his late thirties now, a full professor, respected in his field in the way that careful competent people are respected — not famous, not controversial, not the kind of scientist whose name appears in newspapers, but known among the people who matter in astrophysics as someone whose work can be trusted.

It is 1948. He is thirty-eight years old. He has a career and a department and a life that is proceeding largely as expected. The Cold War has begun and the country is nervous in ways it is still learning to name and the newspapers are full of strange stories from the western states — lights in the sky, objects moving in ways that aircraft do not move, Air Force pilots filing reports that their superiors find inconvenient. The Air Force, which has an interest in knowing whether these objects are Soviet technology and a separate interest in not appearing to take seriously the possibility that they might be something else entirely, has decided it needs a scientist. Specifically it needs an astronomer — someone who can look at a report from a pilot who saw something bright and fast and explain it in terms of celestial mechanics or atmospheric optics or any of the other respectable categories that allow the thing to be classified and filed and forgotten.

Someone gives them Hynek's name.

He will say yes. Of course he will say yes. It is a consulting contract from a respectable institution. It fits his expertise. It will not take much time. There is no reason to say no.

He has absolutely no idea what he is agreeing to.

Chapter Two

The Astronomer

To understand what happens to J. Allen Hynek you have to understand what he was before it happened, and what he was before it happened was a scientist of a specific and serious kind — not a theorist, not a visionary, but a measurer. A man whose relationship with the universe was conducted through instruments, and whose deepest professional satisfaction came from the moment when a careful measurement resolved a question that had resisted less careful ones.

The universe, in Hynek's working life, is not primarily a source of wonder. Wonder is for amateurs and planetarium narrators and the covers of popular magazines. For the working astronomer the universe is a set of problems, and the problems yield, when they yield at all, to patience and precision and the slow accumulation of data taken under conditions rigorous enough that someone else, somewhere else, with the same instruments and the same methods, would get the same numbers. This is not a diminished relationship with the cosmos. It is an intimate one. You do not come to know something deeply by standing back and admiring it.

His particular corner of astrophysics is stellar photometry — the precise measurement of starlight. The question sounds simple and is not. How bright is that star? Not approximately, not impressionistically, but exactly, in a way that can be compared to how bright it was last month and last year and how bright it will be next month, expressed in units that mean the same thing in Columbus as they do in Cambridge or Cape Town. The answer matters because brightness tells you things. It tells you about temperature and composition and distance and age. And for the stars Hynek cares most about — the variables, the ones that brighten and dim on timescales of days or weeks or months — it tells you something even more interesting, which is that stars are not the fixed and permanent objects that the naked eye suggests. They are dynamic, complex, subject to internal forces that produce changes a careful observer can detect and measure and eventually explain.

Variable stars are, in the nineteen-thirties and forties, a genuinely productive field. The mechanisms driving different types of variability are still being worked out. There are the Cepheids, whose regular pulsations have already been used to measure distances to other galaxies, and there are the less well-behaved variables — the irregular ones, the catastrophic ones, the stars that flare or fade in ways that do not fit the existing categories and therefore require new ones. Hynek is drawn to the problem the way careful minds are always drawn to problems that are more complicated than they first appear.

What this requires practically is observation. Sustained, systematic, patient observation over timescales long enough to reveal patterns that short runs of data would miss. You sit at the telescope, or more precisely you sit at the instrument attached to the telescope, and you take measurements, and you record them with sufficient precision that they can be compared against measurements taken months or years later by the same observer or by others. The data accumulates slowly. The patterns emerge from the accumulation. The interpretations follow the patterns. This is not dramatic work. It has the rhythm of farming — the long investment of attention before anything is ready to harvest.

Hynek is good at this. More than good. He has the specific temperament that serious observational science requires, which is not merely patience but a kind of active patience — the ability to remain genuinely engaged with the problem through the long stretches when the data is coming in but the answer isn't yet visible. Lesser scientists become mechanical during those stretches, going through the motions of observation without the quality of attention that makes the observations useful. Hynek does not become mechanical. He stays interested because he has understood something that not all scientists understand, which is that the most important thing you can bring to an instrument is not technique but genuine curiosity about what it is going to tell you.

At Ohio State he builds something. Not just a career — an operation. He improves the photometric equipment, modernizes the methods, trains students in the precision his own training demanded of him. The observatory is not one of the great ones — it does not have the aperture or the sky access of Mount Wilson or Palomar — but under Hynek it becomes a place where serious work gets done, where the data is trustworthy and the methods are sound and the students learn that there is no shortcut between the observation and the conclusion. He is a demanding teacher in the way that people are demanding when they genuinely care about the thing they are teaching — not punitive, not contemptuous of error, but unwilling to accept sloppiness because sloppiness is the enemy of the only thing that matters, which is knowing what is actually true.

His papers from this period are what papers from this period look like — careful, specific, narrow in their claims, building incrementally on what came before. He is not trying to revolutionize the field. He is trying to add real knowledge to it, which is a different and in some ways harder ambition, because it requires resisting the temptation to claim more than the data supports. The temptation is always there. Science has its social pressures like any human enterprise, and the pressure to produce results that are significant, that attract attention, that justify the funding and the time, is constant and in some people corrosive. Hynek is not corroded by it. He has a native resistance to overclaiming that will serve him for the rest of his life, even when the rest of his life takes him somewhere he cannot currently imagine.

The wartime work at Johns Hopkins introduced him to a different mode of scientific practice — applied, urgent, collaborative, oriented toward a specific practical outcome rather than toward the open-ended accumulation of knowledge. The proximity fuze is a genuine technical achievement, one of the significant innovations of the war, and Hynek's contribution to it demonstrates that he can operate in that mode without losing the methodological rigor his academic training built in him. The two modes are not as different as they might appear. Both require careful measurement, honest assessment of what the data says, and the discipline to report results accurately even when the results are not what you hoped for. The difference is that in applied science someone is waiting for the answer and the answer has consequences that extend beyond the scientific community.

Hynek handles this well. He is useful in the way that rigorous people are useful when the stakes are high.

Back at Ohio State after the war, he picks up where he left off and also moves forward. His interests broaden somewhat — he becomes involved in questions of stellar spectral classification, in the instrumentation problems that limit what ground-based astronomy can achieve, in the emerging possibilities of electronic imaging as a replacement for photographic plates. He is not a pioneer in the sense of working alone at the frontier of a new field — he is more the kind of scientist who sees where the frontier is moving and positions himself and his students to contribute usefully to the work that needs doing there. This is a respectable and genuinely valuable role in science, less celebrated than the pioneer's role but equally necessary. Someone has to do the careful secondary work that tests the pioneers' claims and refines their methods and catches their errors. Hynek does some of this and is good at it.

His worldview in this period is what his training and his temperament have made it. He is a positivist in the philosophical sense — he believes that knowledge comes from observation and experiment, that claims not grounded in measurable evidence are not scientific claims, that the history of human understanding is the history of replacing speculation with data. He is not dogmatic about this — he is too careful a thinker to be dogmatic about anything — but it is his default orientation, the lens through which he looks at questions about the natural world. If you cannot measure it, if you cannot test it, if you cannot in principle be shown to be wrong about it, then it is not science. This is not a narrow or dismissive view. It is the view that has produced every genuine advance in human understanding of how the universe works, and Hynek holds it with the quiet confidence of someone who has seen it vindicated repeatedly in his own work.

He has, within this worldview, a specific and considered contempt for a certain kind of irrationality — the kind that presents itself as science without doing the work that science requires. Astrology irritates him, not because the stars are not interesting but because astrology makes claims about the stars without the measurements and the rigor and the accountability to evidence that make claims about the stars meaningful. Paranormal claims of various kinds produce in him a mild professional disdain — not hostility, because he is not a hostile man, but the particular impatience of someone who has spent his career learning how hard it is to know anything reliably, watching other people claim to know things they have not earned the right to claim.

This is worth establishing carefully, because it means that when the Air Force consulting work begins he is not secretly open to the possibility of something unexplained. He is not, underneath the professional skepticism, a romantic waiting to be validated. He is a working scientist who has organized his entire intellectual life around the principle that extraordinary claims require extraordinary evidence, and who has, in thirty-eight years, encountered nothing that has given him serious reason to revise that principle.

He is, in other words, exactly the wrong person for what is about to happen to him.

Or perhaps exactly the right one.

He moves to Northwestern University in 1960, where he will chair the astronomy department and have access to the Dearborn Observatory on campus, a nineteenth-century instrument that is more historically significant than scientifically competitive but that suits the teaching and outreach work that becomes an increasing part of his role. He is fifty years old, established, respected, a fixture of midwestern academic astronomy. He has graduate students who will go on to productive careers. He has colleagues who trust his judgment. He has a life organized around the careful pursuit of knowledge about things that are very far away and can be measured with the right instruments from a sufficient distance.

The consulting work for the Air Force has been running in the background for twelve years by now, a minor irritant, occasionally interesting, more often tedious, never the kind of thing he would mention first if someone asked him what he did. He has explained away hundreds of sightings. He has provided the astronomical context that allows strange reports to be filed and forgotten. He has been useful to an institution that needed someone with his credentials to perform a function that his credentials made possible.

He has not, in twelve years, found a reason to take the subject seriously.

He has also not found a reason to stop.

Chapter Three

The Consulting Gig

The letter, or the phone call, or however it arrives, comes to him in 1948 and does not change his life immediately. That is the thing about the moments that change lives — they rarely announce themselves as such. They arrive in the ordinary packaging of ordinary requests and the person receiving them says yes or no based on the information available at the time, which is never the information that would have mattered. Hynek says yes. Of course he says yes. It is a reasonable request from a respectable institution and it fits his expertise and it will not, he is assured, take very much of his time.

The request comes from the United States Air Force, which in 1948 is a young institution — it has existed as a separate branch of the military for less than a year, having been carved out of the Army in September 1947 — and which has a problem it is not sure how to characterize. Since June of the previous year, when a civilian pilot named Kenneth Arnold reported seeing nine crescent-shaped objects flying in formation near Mount Rainier at speeds he estimated at nearly seventeen hundred miles an hour, the reports have not stopped coming. Pilots, soldiers, civilians, people with every reason to be believed and people with rather less — all of them filing accounts of objects in the sky that do not correspond to any aircraft in the American inventory and do not behave in ways that known aircraft can behave. The Air Force has two concerns and they exist in tension with each other. The first is that some of these objects might be Soviet technology of a kind that would represent a serious intelligence failure. The second is that publicly acknowledging uncertainty about objects in American airspace would undermine the confidence of a population already nervous about what the Cold War might bring.

What they need, they have decided, is a scientist. Specifically an astronomer, because a significant number of the reports describe things in the sky and an astronomer is the credentialed authority on things in the sky and an astronomer can say, with the weight of professional expertise behind the statement, that what the witness saw was a meteor, or a weather balloon, or Venus in an unusual atmospheric condition, or any of the other natural phenomena that naive observers regularly misidentify. The explanation does not need to be exciting. It needs to be plausible and it needs to come from someone whose credentials make it difficult to argue with.

Someone at Wright-Patterson Air Force Base in Dayton, Ohio, which is where the project is headquartered, thinks of Ohio State University, which is nearby, and Ohio State University's astronomy department, which has a professor named Hynek whose work they probably cannot describe in detail but whose institutional affiliation gives him the right kind of standing. [*Verify: specific recruitment pathway and who made initial contact.*] He is thirty-eight years old. He has a doctorate and a faculty position and a modest but genuine reputation in stellar photometry. He is not famous. He is not controversial. He is, from the Air Force's perspective, exactly what they are looking for — a credentialed scientist willing to do consultancy work for a government client, close enough to Dayton to be convenient, experienced enough to be useful, and sufficiently removed from the subject matter to approach it without the kind of prior investment that would complicate his conclusions.

Hynek's own reaction to the invitation is not recorded in detail, but its broad shape is easy to reconstruct from what he later wrote about his early attitude toward the phenomenon. He thinks the subject is, to put it plainly, nonsense. The flying saucer craze of 1947 has produced, in his professional assessment, exactly what you would expect a public craze to produce — a wave of misidentified conventional objects, atmospheric phenomena, and the occasional outright fabrication, all of it amplified by newspapers that have discovered a story their readers cannot get enough of. He does not believe there is anything genuinely unknown in these reports. He believes that a competent astronomer applying standard methods of analysis will be able to explain the significant majority of cases quickly, and that whatever small residue remains after that analysis will dissolve upon closer examination. He is not dismissive in a contemptuous way — he is dismissive in the way of someone who has a calibrated sense of how often extraordinary claims turn out to have ordinary explanations, which in his experience is nearly always.

He takes the contract.

The project he is joining is called Project Sign, and it is the Air Force's first formal attempt to investigate the reports systematically. It is headquartered at Wright-Patterson's Air Materiel Command and staffed by a mixture of military officers and civilian analysts who have varying levels of enthusiasm for the work and varying views about what the work is likely to find. Some of them take the reports more seriously than their official posture allows them to say publicly. Others are performing the kind of institutional due diligence that exists to demonstrate that a potential problem has been examined rather than to actually examine it. The atmosphere is, from the beginning, uncomfortable with genuine inquiry — an organization that has been asked to investigate something it has already, at some level, decided the answer to.

Hynek's role is defined clearly enough. He receives reports — summaries of sighting accounts filed by military witnesses, civilian reports forwarded through official channels, occasionally the underlying raw documentation from a particularly significant case — and he provides astronomical analysis. Is this consistent with a meteor? Could this be the planet Venus under these atmospheric conditions? Is there a known satellite or high-altitude balloon program that could account for this description? He is not making policy. He is not running investigations. He is providing the scientific context that allows cases to be categorized and closed, and he is doing this from Ohio State, at a comfortable remove from the institutional pressures of Wright-Patterson, using the tools and methods he has spent his career developing.

He is, at this, genuinely competent. This is worth saying because it would be easy, in retrospect, to characterize his early debunking work as mere compliance — a scientist performing a function he knows to be dishonest in exchange for a government paycheck. That is not what it is, at least not in the beginning. When Hynek says that a sighting is consistent with the planet Venus, he means it. When he identifies a reported object as a probable meteor on the basis of the witness description of its trajectory and duration, the identification is real. The majority of reports that cross his desk in 1948 and 1949 do yield to this kind of analysis, because the majority of reports are what he expected them to be — misidentifications of conventional objects by observers who lack the astronomical training to recognize them. He is providing a genuine scientific service and he knows it.

The problem, which does not announce itself as a problem immediately, is the residuals.

Any analyst processing a large dataset expects a residue — the percentage of cases that resist the initial categorization, that don't fit the available explanations, that require either further investigation or the acknowledgment that the explanation is not yet available. In most scientific contexts this residue is interesting but not alarming. You note it, you continue collecting data, you eventually either find the explanation or you revise your model to account for the anomaly. Hynek expects a small residue in the Blue Book cases. What he does not quite expect — what he notices but does not yet name — is the character of the residual cases. They are not random noise. They are not the reports from obviously unreliable witnesses or the accounts so vague that no analysis is possible. They tend, with a consistency that begins to register somewhere in the back of his careful mind, to be the reports from the most credible witnesses, described with the most precision, containing the most specific and verifiable detail. The cases that should be easiest to explain, because the witnesses are trained observers filing careful accounts, turn out to be the cases that resist explanation most stubbornly.

He files this observation and moves on. He is busy. He has his actual career to attend to, his students and his papers and his observatory, and the Blue Book work arrives in batches and gets processed and returned and the checks arrive and life continues. He is not lying awake thinking about the residuals. He is a scientist with a full professional life who has a part-time consulting arrangement that produces more tedium than mystery. The observation goes into the mental file where observations that don't yet have a home are stored, and the file sits there, and Hynek goes back to his variable stars.

Project Sign produces, in 1948, a document that will not become public for years but that represents a significant moment in the institutional history of the phenomenon. Known as the Estimate of the Situation, it is an internal Air Force assessment that concludes, on the basis of the evidence accumulated in the project's first year of operation, that the most credible unexplained reports are consistent with the hypothesis that the objects are interplanetary in origin. This is a remarkable document for a military institution to produce, and it does not survive the institutional review process — General Hoyt Vandenberg, the Air Force Chief of Staff, rejects it, reportedly on the grounds that the evidence is insufficient, and orders the copies destroyed. What this means in practice is that the project's own internal analysis, conducted by the people most familiar with the data, reaches a conclusion that the institution is not prepared to accept, and the institution responds by suppressing the conclusion rather than engaging with the evidence that produced it. Sign is reorganized and renamed Project Grudge, and the atmosphere of the project shifts perceptibly toward the conclusion that has been decided upon rather than the conclusion the data might support.

Hynek is a peripheral figure in this institutional drama — he is a consultant, not a staff member, and the internal politics of Wright-Patterson are not his daily concern. But he is aware of what has happened, at least in outline, and he finds it professionally uncomfortable in the way that any scientist finds it uncomfortable when institutional conclusions precede institutional evidence. He does not make a fuss about it. It is not his project and not his fight, and the conclusion the Air Force has reached — that the reports are misidentifications and hoaxes and the residuals are simply unexplained rather than inexplicable — is not, in 1949, so far from his own working assumption that he feels compelled to object. He files another observation in the mental file and continues.

This is how the accommodation works, in the beginning. Not through corruption or cowardice but through a series of individually reasonable decisions that collectively produce an unreasonable result. Each step is small enough to be defensible. He takes the contract because it is reasonable. He provides the analyses because they are genuine. He files the anomalies because he has insufficient data to act on them. He does not object to the institutional politics because it is not his battle and the institutional conclusions are not yet far enough from his own to demand objection. He is not being dishonest. He is managing the situation the way a careful professional manages situations that are imperfect but not yet intolerable.

The file in the back of his mind is getting thicker.

Project Grudge runs from 1949 to 1951, operating with considerably less rigor than Sign and with a mandate that amounts, in practice, to finding conventional explanations for reports regardless of whether the evidence supports those explanations. The Air Force, having decided that the phenomenon is not a national security threat and not worth the public relations complications of genuine investigation, wants the cases closed efficiently. Hynek continues his consulting role, continues providing astronomical analysis, continues doing real work on the cases that his expertise is genuinely relevant to. He is a professional. He delivers what he is asked for.

But something is happening underneath the professional surface that Hynek himself probably cannot fully articulate in 1951. He has now processed hundreds of cases. He has applied his methods carefully and honestly to each one. He has found conventional explanations for the majority and has said so, honestly. And the residuals — the cases that his honest application of careful methods cannot explain — have not gone away. They are still there. They are still, characteristically, the best-documented cases from the most reliable witnesses. They are accumulating in the mental file with the quiet persistence of data that does not care whether it is convenient.

He is still, in 1951, a skeptic. He still believes, or tells himself he believes, that the residuals will eventually yield to better analysis or additional data or some conventional explanation not yet identified. He is not ready to say otherwise and would not say otherwise if asked. The professional distance is intact. The contract is intact. The life at Ohio State is intact and proceeding normally and the consulting work is a minor irritant at the margin of a career organized around other things.

But the file is there.

And it is getting harder to ignore.

Chapter Four

The Company Man

Project Blue Book begins in 1952 and will run for seventeen years, which is a long time to do anything and an especially long time to do something that requires you to maintain a position you are increasingly uncertain you hold. Hynek does not know in 1952 that he will still be consulting for the project in 1969. He does not know that Blue Book will become the defining institutional experience of his professional life, longer than any single appointment, longer than any research program, outlasting colleagues and administrations and his own certainty about what the cases mean. He knows only that the contract continues and the cases keep arriving and his analyses are still, in the main, producing conventional explanations for unconventional reports, and that the residuals are still there.

The project has a new name and nominally a new orientation. Where Grudge had operated with the transparent mandate of dismissal, Blue Book is structured to suggest genuine inquiry — a case numbering system, standardized report forms, a small permanent staff at Wright-Patterson dedicated to processing the incoming material. Its first director is Captain Edward Ruppelt, a young Air Force officer who is, by the accounts of everyone who worked with him, one of the more intellectually honest people ever to run the project. Ruppelt actually wants to know what the cases represent. He approaches the material with something resembling genuine scientific curiosity, and he recruits and consults people — Hynek among them — whom he respects as analysts rather than merely as credentialed cover.

The Ruppelt years are, for Hynek, the most professionally satisfying period of his Blue Book tenure, for the simple reason that Ruppelt does not tell him what conclusions to reach. He asks questions and waits for answers and treats the answers as information rather than as problems to be managed. Hynek can do his work honestly under these conditions. He takes cases seriously. He visits sites when the reports warrant it. He talks to witnesses with the careful attention he brings to any observational problem — listening not just for what they say but for how they say it, for the specific sensory details that separate genuine experience from confabulation, for the internal consistency that distinguishes a real account from a constructed one. He is developing, without quite naming it as such, a methodology for evaluating witness testimony that draws on his scientific training but extends beyond it into territory that his training did not cover.

He also begins, in the Ruppelt years, to develop a typology — an informal mental classification of the kinds of reports that cross his desk. Most cases are what he expects: misidentified aircraft, weather balloons, astronomical objects seen under unusual conditions. A smaller category are the cases he privately labels as unknowns — not because he is prepared to say they represent something genuinely anomalous, but because his honest application of the available analytical tools does not produce a satisfying conventional explanation. And within the unknowns there is a smaller subset still, cases that he finds himself returning to, that resist not just his first analysis but his second and third, that have a quality of detail and credibility that makes the conventional explanation feel not merely uncertain but actively wrong.

He does not say this to anyone at Wright-Patterson. He writes it in his private notes in the careful, qualified language of a scientist who knows that private uncertainty is not public evidence. The professional distance, still intact, still serving its function, allows him to hold the private thought and the public position simultaneously without experiencing them as a contradiction. He is not hiding something he knows to be true. He is managing something he suspects might be true but cannot yet demonstrate, which is a different and more defensible position, and one that a rigorous scientist can occupy for quite a long time before it becomes untenable.

Ruppelt leaves Blue Book in 1953 and the project changes. His successors do not share his orientation toward genuine inquiry, and the institutional pressure that Ruppelt had held at a manageable distance reasserts itself. The mandate, never explicitly stated but impossible to misread, is efficiency — close cases, produce conventional explanations, keep the residual percentage low, avoid generating material that might interest journalists or congressmen or members of the public who have decided that the government is hiding something. The atmosphere Hynek returns to on his periodic visits to Wright-Patterson is different from the one Ruppelt created — more defensive, more oriented toward outcome, less comfortable with the kind of genuine analytical uncertainty that honest investigation requires.

He adapts. This is not a flattering thing to say about a man of genuine integrity, but it is accurate, and the biography that omits it in favor of a cleaner portrait does Hynek a disservice. He adapts because the alternative — refusing the contract, making a public issue of the institutional dishonesty, sacrificing the consulting income and the access to the case files — requires a certainty he does not yet have. He cannot say publicly that Blue Book is suppressing evidence of something genuinely anomalous because he cannot yet say privately that he is certain there is something genuinely anomalous to suppress. He has suspicions. He has a thickening file of cases that trouble him. He does not have proof, and a scientist who has organized his entire intellectual life around the principle that claims require evidence is not going to make a public claim he cannot substantiate, even when the institutional behavior around him is making him increasingly uncomfortable.

What this produces, across the middle years of the nineteen-fifties, is a man performing a function he has mixed feelings about, at a level of quality he still, mostly, maintains, while the mixed feelings gradually acquire more weight. He provides astronomical analyses. He visits sites when asked. He talks to witnesses. He writes reports. The reports, increasingly, find conventional explanations for cases that his private notes suggest he finds more ambiguous than the reports indicate. The gap between the public document and the private assessment is not large in any individual case — he is not fabricating explanations he considers obviously false — but it is consistent, and consistency across enough cases produces something that begins to resemble a pattern of systematic understatement.

He is becoming, without fully acknowledging it, a man who is telling partial truths for an institution that requires partial truths. Not lies exactly. Possibilities offered as probabilities. Hypotheses treated as conclusions. The range of uncertainty in his assessments narrowed for public consumption beyond what his actual confidence in those assessments would justify. He is still a rigorous scientist. The rigor is still there. But it is being applied selectively, in the direction the institution requires, and the selectivity is doing something to him that will take years to fully register.

The Robertson Panel convenes in January 1953, before Ruppelt's departure, and produces a document that Hynek will later characterize as one of the most damaging developments in the history of serious UFO investigation. The panel — assembled by the CIA and chaired by the physicist H.P. Robertson of Cal Tech — reviews a selection of Blue Book cases over four days and concludes that the reports represent no national security threat. More significantly, it recommends that the Air Force actively debunk public interest in the subject to prevent the communications channels needed in a genuine emergency from being clogged with UFO reports. Hynek attends the panel as an associate member. He reads a prepared paper to the panel, briefs them on the Battelle Memorial Institute study of Blue Book cases, and participates in discussions. He is not a peripheral figure — he is present and engaged. And he is quietly resistant. Ruppelt later notes that Hynek and the astronomer Thornton Page tend to line up against the more dismissive members, asking harder questions, pushing back on the speed of the conclusions. His discomfort is on the record, if anyone is looking.

It is a conclusion reached in four days about a phenomenon that Hynek has been examining for five years, by a panel that reviewed a curated selection of cases rather than the full body of data, and it recommends, explicitly, that the government use media and public education programs to reduce public interest in something that Hynek has in his private file considerable evidence suggesting deserves more interest rather than less. He does not resign. He does not publicly object. But he writes in his notes something that suggests he finds the panel's methodology inadequate — four days is not enough time, the case selection is not representative, the conclusion precedes rather than follows the analysis. He continues. The contract continues.

The cases accumulate. In 1957 there is a wave of reports from across the United States that produces more Blue Book case numbers in a shorter period than anything since 1952, driven partly by Soviet Sputnik anxiety — Hynek is at the Smithsonian Astrophysical Observatory by this point, running Operation Moonwatch, the satellite tracking program for the International Geophysical Year, and is appearing on national television and on the cover of LIFE magazine in the weeks after Sputnik's launch. He is, for the first time in his life, a public figure of a kind, and the public figure is an astronomer of conventional respectability. The UFO consulting work runs in parallel, invisible to the audience watching him discuss satellite orbits on the evening news, invisible to his colleagues at the Smithsonian who respect him for his role in tracking objects that everyone agrees are actually there.

The 1957 sighting wave produces, across unconnected reports filed by observers with no knowledge of each other, a consistency of description that Hynek finds increasingly difficult to account for through the standard mechanisms of mass misidentification. He processes the cases. He writes the analyses. He closes the files where they can be closed and marks the residuals as unidentified and moves on. But the residuals are no longer a statistical curiosity at the margin of an otherwise explicable dataset. They are a pattern. And a pattern, in science, demands explanation.

This is the company man in full operation — not corrupt, not cynical, but accommodating. Telling himself that his continued involvement allows him to do better work than his absence would permit. Telling himself that the residuals will eventually accumulate to a point where the evidence is undeniable and the public statement will be possible. Telling himself that he is still a scientist doing scientific work and that the institutional context, however imperfect, does not fundamentally compromise the integrity of the analysis. These are the stories that people tell themselves when they are doing something they are not entirely comfortable with, and Hynek is smart enough that his versions of them are more sophisticated than most. He is also honest enough that they are not, in the deepest part of him, entirely convincing.

He moves to Northwestern University in 1960, leaving Ohio State, and takes up the chairmanship of the astronomy department. He is fifty years old, established, respected, a fixture of midwestern academic astronomy. He has graduate students who will go on to productive careers. He has colleagues who trust his judgment. The consulting work for the Air Force has been running in the background for twelve years. He has explained away hundreds of sightings. He has been useful to an institution that needed someone with his credentials to perform a function that his credentials made possible.

Major Hector Quintanilla takes over Blue Book in 1963 and will run it until its closure in 1969. Quintanilla considers Hynek a liability. The word is documented, and it is exactly the right word for what Hynek has become from the Air Force's perspective — a consultant who keeps asking questions the institution does not want asked, who pushes back in meetings, who writes private analyses that diverge from the public conclusions, who cannot quite be trusted to perform the function he was originally recruited to perform without the complication of his actual scientific judgment getting in the way. The institution needs a debunker. It has ended up with something more inconvenient: a rigorous scientist who is not convinced by his own debunking.

There are nights, probably, when he takes out the private file — the cases that resist explanation, the witnesses whose accounts he cannot dismiss, the patterns that appear across reports filed by people with no connection to each other — and sits with it in the way that a scientist sits with a problem that will not resolve. He is trained to sit with unresolved problems. He is trained to hold the question open until the evidence is sufficient to close it. But there is a difference between a question about the brightness of a variable star and a question about whether the Air Force is systematically suppressing evidence of something that the public has a right to know about, and the discipline that serves him in the first case does not map cleanly onto the second.

He is not ready. He is not sure he will ever be ready. He gets back on the plane to Dayton and processes the next batch of cases and writes the reports and cashes the check and goes back to his stars.

But something has changed in him that cannot be unchanged. He knows things, in the private file, that the public version of J. Allen Hynek does not know. He has seen cases he cannot explain. He has talked to witnesses he cannot dismiss. He has watched an institution choose conclusion over evidence in ways that a scientist is trained to recognize as a corruption of the process that produces reliable knowledge.

He is still telling himself it is fine.

He is telling himself that less convincingly every year.

Chapter Five

The Seduction

There is a specific kind of scientific attention that is different from ordinary professional engagement, and Hynek has been trying for years not to give it to the Blue Book cases. Ordinary professional engagement is what he has been practicing: take the file, apply the methods, produce the analysis, close the case or mark it unidentified, move on. The cases are processed like data because they are data, and a scientist processes data without becoming personally involved in it. That is the discipline. That is what has allowed him to do this work for fifteen years without it consuming him.

The other kind of attention — the kind he has been trying to avoid — is the attention that a scientist gives to a problem that will not resolve. It is active, hungry, recursive. It keeps returning to the anomaly not because duty requires it but because something in the mind will not let the unresolved thing rest. It is, in the most precise sense, fascination: the state of being held by something against your will. Hynek has been fending off fascination for a long time. He is very good at it. He is about to discover that he cannot do it forever.

The year is 1963. He is at Northwestern, chairing the astronomy department, and a young Frenchman named Jacques Vallée arrives as a graduate student. Vallée has come from the University of Texas at Austin, where he was working on astronomical mapping for NASA, and before that from the Paris Observatory, where, as a young staff astronomer on the French Space Committee, he had watched a superior order the destruction of tracking tapes recording unknown objects in earth orbit — an act of institutional suppression that had lodged in him an interest in the UFO phenomenon that his scientific training had not extinguished but had instead made more rigorous. He is twenty-four years old, precise, intellectually fearless, and in possession of a mind that processes patterns across large datasets with the ease of someone for whom that kind of thinking is simply natural. He will eventually earn his doctorate from Northwestern in computer science. He is already thinking about building the first computerized database of UFO sightings.

Hynek recognizes him immediately — not personally but categorically. Here is someone who sees what he sees in the Blue Book data and is not frightened by it, who applies the same rigorous analytical habits to the phenomenon that their shared discipline demands of any problem, and who has not yet spent fifteen years learning to manage his conclusions for institutional consumption. The relief of that recognition is probably considerable. Hynek has been, in a specific sense, alone with this material for a long time. He has his private file and his private doubts and the careful professional distance he maintains in his public analyses, and no one in his department or his field with whom he can think honestly about what the data actually says. Vallée gives him that. Their collaboration begins quickly and becomes close.

The term Hynek uses for what they are building together, along with a handful of other credentialed scientists who are privately asking the same questions from their own institutional positions, is the invisible college. He borrows it from the seventeenth-century network of natural philosophers who exchanged observations and ideas outside the formal structures of the learned societies, before those structures existed, creating an unofficial community of inquiry that operated in the gaps between what could be said publicly and what was actually being thought. The parallel is deliberate and precise. There are scientists in aerospace, in physics, in psychology, in atmospheric optics, who have encountered the UFO data in one form or another and found it troubling and cannot say so in their professional lives without consequences they are not willing to face. They are not cranks. They are not enthusiasts. They are careful people who have looked at careful evidence and reached careful tentative conclusions that their institutions do not want them to reach. Hynek knows several of them. He begins, cautiously, to connect them.

The invisible college is both a practical arrangement and a symptom. Practically, it allows the sharing of data and analysis that cannot happen through official channels without triggering the institutional suppression that the Robertson Panel formalized a decade earlier. As a symptom, it reveals the size of the gap between what the scientific community is privately thinking about the phenomenon and what it is publicly willing to say. Hynek is at the center of this network by virtue of his Blue Book access, which gives him a dataset that none of the others can match. But the network also tells him something about that dataset — that the patterns he has been seeing in the Blue Book files, the patterns he has been carefully filing away for fifteen years without acting on them, are visible to other careful minds looking at smaller samples of the same material. He is not seeing things that are not there. Other people are seeing them too.

Then comes Socorro.

On the afternoon of April 24, 1964, Lonnie Zamora, a Socorro, New Mexico police officer on routine patrol, is chasing a speeding vehicle south of town when he hears a roar and sees a flame descending in the southwestern sky. He breaks off the chase, heads toward the sound, and crests a hill to find, in the arroyo below, a small egg-shaped object standing on four legs in the desert scrub. Near it, two figures in white coveralls. At fifty feet Zamora stops, gets out of his cruiser, and looks. The object displays a red insignia on its side — an inverted V with a horizontal bar and three lines crossing it, which Zamora later draws for investigators in careful detail. Then the roar begins again, the figures disappear, and the object lifts from the desert on a blue and orange flame, rises to a low altitude, and departs to the southwest in silence.

Zamora is thirty-one years old. He has been a police officer in Socorro for a decade. He reports what he has seen with the clipped, sequential precision of a law enforcement professional describing an incident for the record, without drama, without elaboration, without the kind of interpretive overlay that a witness with a prior investment in the subject tends to produce. The burn marks and landing gear impressions in the arroyo floor are documented by multiple investigators before they degrade. Five tourists driving through Socorro at roughly the right time later report seeing something bright rise from near the town. A master sergeant at White Sands, south of the incident site, reports a strange blue and orange light coinciding with vehicle interference the same evening.

Hynek is ordered by Quintanilla to investigate within days of the incident, and does so immediately. He interviews Zamora. He examines the site. He conducts a follow-up investigation in August 1964. He tries, with the thoroughness his training demands, to find the conventional explanation — the military test, the student hoax, the misidentification — and cannot. In correspondence with the astronomer Donald Menzel, who proposes an elaborate hoax hypothesis involving a balloon, Hynek presents the theory directly to Zamora, who rejects it flatly with the unmoved confidence of a man who knows what he saw. Writing to Menzel in September 1964, Hynek allows himself a formulation that his public role has never previously permitted him: this case, he writes, may be the Rosetta Stone, and Zamora is the most unimpeachable witness he has ever encountered in sixteen years of investigation. That phrase — most unimpeachable witness — is the private measure of the gap Hynek has been managing between what he knows and what he says. He has never said anything like it before, even in private. He says it now because the evidence for Socorro is, by his own rigorous assessment, simply the best he has ever seen.

Blue Book lists the Socorro case as Unknown. Quintanilla, in a CIA publication two years later, writes that there is no doubt Zamora saw something that left an impression on him, but declines to speculate further. The case is, in Quintanilla's own assessment, one of the best documented on record. It is also, in Quintanilla's practice, a case to be acknowledged and set aside rather than pursued. The institutional machinery for turning good evidence into non-evidence operates exactly as it always has. File it as Unknown, which means neither explained nor investigated further, and wait for the public attention to move on.

Hynek does not move on. He cannot, quite. The Socorro case does something to him that the accumulated residuals of sixteen years have been building toward but have not quite achieved — it makes the phenomenon personal. Not in the sense of an emotional response to a dramatic event, but in the precise sense that a problem becomes personal to a scientist when the evidence crosses the threshold where professional courtesy to conventional explanation is no longer intellectually honest. He has reached that threshold before, privately, with individual cases. Socorro is the case where he reaches it and knows he has reached it and cannot unknow it.

What follows, through 1964 and 1965, is a period of intensifying private conviction running alongside a public posture that has not yet caught up with it. The invisible college becomes more active. Vallée is building the computerized database, and the patterns emerging from systematic analysis of large numbers of cases are confirming what Hynek has observed impressionistically for years — the residuals are not random, they are structured, they have characteristics that repeat across time and geography in ways that misidentification and fabrication cannot account for. Physical trace evidence appears in multiple independent cases. Electromagnetic interference associated with close approaches is documented across unconnected reports. The cases cluster in ways that suggest, to a mind trained to look for signal in noise, something other than coincidence.

Hynek is also, in this period, observing the institutional response to good evidence up close and finding it increasingly difficult to defend on any scientific grounds. When he tries to encourage a proper search for additional witnesses to the Socorro incident, he is told by Blue Book personnel that this generates no interest whatsoever. When cases with strong witness testimony and physical evidence land on his desk, he watches Quintanilla's staff find explanations for them with a speed and confidence that bears no relationship to the quality of the analysis. The gap between what the evidence warrants and what the institution produces is no longer something he can explain as reasonable caution or analytical disagreement. It has become, unmistakably, policy.

He is still consulting. He is still writing the analyses. He is still performing the function he was recruited to perform in 1948. But the man performing that function is no longer the man who was recruited to perform it. The astronomer who took a modest consulting contract because it fit his expertise and would not take much time has been, case by case, year by year, replaced by someone who knows things that cannot be unknowingly replaced. The seduction that he resisted for sixteen years has accomplished what seduction always eventually accomplishes when the thing doing the seducing is real: it has made him care.

He is careful not to show it. At Wright-Patterson he is still the professional consultant, still the astronomer providing qualified analyses within the framework the institution requires. In the invisible college he is more candid, more willing to say what the evidence suggests without the hedges that his public role demands. In his conversations with Vallée, which Vallée is recording in the journals he will later publish, he is something else again — a scientist thinking out loud at the edge of what his training and his honesty will jointly allow him to say, pushing at the boundary between the defensible conclusion and the conclusion the evidence actually supports, not yet willing to cross it publicly but no longer pretending the boundary is as far from crossing as he has spent sixteen years insisting it is.

It is late 1965. Project Blue Book has been running for thirteen years under his consultancy. He has processed thousands of cases. He has provided explanations for the large majority and marked the rest unidentified. The unidentified cases sit in the files at Wright-Patterson and do not trouble the institution, because the institution has made its peace with unidentified cases — they represent the irreducible noise in any large dataset, the tail that will always escape the net, the cost of doing business. This is the institutional understanding of the residuals, and it has been Hynek's official understanding as well, the position he has defended in briefings and analyses and public appearances for over a decade.

It is no longer his private understanding. Privately, he thinks the residuals are the story. He thinks the cases that resist every conventional explanation are not noise — they are signal. He thinks the patterns that Vallée is finding in the computerized data are real patterns, that the physical evidence from trace cases is real evidence, that witnesses like Lonnie Zamora are telling the truth about experiences that have no place in any established category of natural phenomena. He thinks, though he cannot yet say it in those words in any official context, that something is genuinely happening that science has not yet found the framework to address.

He is fifty-five years old. He is chairman of the astronomy department at Northwestern University. He has a pipe and a beard and a reputation and a consulting contract with the United States Air Force and a private file that contradicts everything his public role requires him to say.

In March 1966, the phones begin ringing in Michigan.

Chapter Six

Swamp Gas

The phones begin ringing in the Washtenaw County Sheriff's Office on the night of March 14, 1966, and do not really stop. For nearly three hours, beginning around 3:50 in the morning, residents across Lima Township report lights moving at speeds that no aircraft in the inventory can match, executing maneuvers that no aircraft in the inventory can execute, and then disappearing. The callers come from Monroe, Livingston, Ypsilanti, Dexter, and Sylvania. Sheriff Douglas Harvey — a practical, skeptical man who has been running this county for years and knows his community well — takes the calls and does not know what to make of them. His own deputies are among the witnesses. He trusts his deputies.

Six days later, on the evening of Sunday, March 20, the phones ring again. Frank Mannor, a truck driver who lives on a farm outside Dexter Township, reports that he and his teenage son have followed something into the swamp behind their house. They got to within five hundred yards of it, Mannor tells the deputies who arrive, and what they saw was shaped roughly like a pyramid, with a blue-green light on the right-hand side and a white light on the left. The body of the thing looked, Mannor says, like yellowish coral rock with holes in it, like if you took a piece of cardboard and split it open. There was no sound they could identify as mechanical. It hovered. Then it was gone. More than forty officers join a search of the swamp before the night is over. The search finds nothing that explains what Mannor described.

The following night, fifty miles southwest at Hillsdale College, the women in MacIntyre Residence Hall are awakened by something moving outside their dormitory windows over the college's Slayton Arboretum. They watch it for hours — a glowing, football-shaped object, according to Blue Book's own subsequent file description, hovering over the low depression of the arboretum and moving in ways that none of them can account for. They call William Van Horn, the Hillsdale County civil defense director, who arrives with a patrol car and watches it himself. By the time it disappears around 5:10 in the morning, eighty-seven people have observed it. They have taken contemporaneous notes. They are not a credulous group — they are college students, most of them, with the particular skeptical intelligence of young people who consider gullibility beneath them, and they are baffled.

Harvey has now reached the limit of what local law enforcement can do with a situation of this kind. He reaches out to several federal agencies and is ignored. He goes to Congressman Weston Vivian, who takes him seriously and pulls the necessary strings. On March 22, papers across the country announce that the Air Force is dispatching its scientific consultant from Northwestern University — J. Allen Hynek, astrophysicist, veteran of eighteen years with Project Blue Book — to investigate the sightings. Hynek arrives the following day.

Harvey drives him to the Mannor farm. They walk the site together. Harvey shows him the circular marks in the vegetation at the reported landing location. He watches Hynek examine the evidence with the careful attention of a man who has done this many times — crouching, looking, measuring with his eyes, saying little. In the car on the way back, Hynek speaks. There is something here, he tells Harvey. They have been investigating this kind of thing for quite a while and they cannot put their finger on it. Harvey, who has now been living with this material for ten days and has staked his professional credibility on the witnesses being honest, finds this reassuring. The government's own scientist is telling him the evidence is real.

Back at the sheriff's office, Hynek asks to use the phone privately. Harvey steps out. Hynek is on the phone for quite a while, Harvey will recall forty years later. When Hynek comes back out, Harvey asks him directly: What do you think, Doctor? Hynek says: It's swamp gas.

Harvey stares at him. He tells me one minute he has no idea what it is, Harvey will say, and then he makes one phone call to Washington and comes out and gives a statement that it's swamp gas. Very strange. Harvey is a law enforcement professional. He knows the difference between a man who has reached a conclusion and a man who has been given one. He does not accuse Hynek of anything. He simply notes, with the precise bewilderment of a witness who has seen something he cannot explain: very strange.

The press conference is scheduled for March 25 at the Detroit Press Club. Sixty reporters attend. Hynek stands at the podium and delivers his explanation: the sightings at Dexter and Hillsdale were the result of marsh gas — methane produced by the decomposition of organic matter in wetlands, sometimes luminous, sometimes mistaken at night for something more dramatic. He explains the phenomenon in technical terms, carefully, the way he would explain anything. He says it is highly localized. He notes, with perhaps more flourish than the situation requires, that a dismal swamp is a most unlikely place for a visit from outer space. He also dismisses the photographs taken by a Washtenaw County police officer as time-exposure trails of the rising crescent moon and the planet Venus.

He knows he is in trouble when he sees a reporter underline the words swamp gas on his notepad and dash for the telephone. He watches it happen in real time — the reduction of a technical explanation to two words that are going to sound, in every headline in the country by morning, like exactly what they are: an insult. The room is not persuaded. Several reporters have stopped writing. Will Van Horn, the Hillsdale civil defense director who watched the object himself for hours, is in the room, and he is not persuaded. The witnesses who drove to the press conference from Dexter are not persuaded. A reporter hands Hynek a magazine showing George Adamski's famous flying saucer photograph. Hynek looks at it and says it looks like a chicken brooder. The remark gets a laugh from some quarters and nothing at all from others. The room is moving in a direction he cannot reverse.

The response is nearly universal and nearly instantaneous. The South Bend Tribune's headline is representative: Air Force Insults Public with Swamp Gas Theory. Dozens of papers accuse the government of suppressing evidence. Witnesses tell reporters they know what swamp gas looks like and this was not it. On the Tonight Show, Johnny Carson interviews a Cal Tech scientist who says the explanation does not fit the facts. Gerald Ford — House Minority Leader, congressman from Michigan's fifth district, future president of the United States — writes a letter to the chairmen of the House Armed Services Committee and the House Science and Astronautics Committee on March 28, describing Hynek's explanation as flippant and calling for a formal congressional investigation. The American public, Ford writes, deserves a better explanation than this. Sheriff Harvey tells the Ann Arbor News: with all due respect to Dr. Hynek, he is not ready to accept this weak excuse of gas from marshes. Walter Cronkite covers the story on CBS Reports. Life Magazine runs a feature in its April issue. Multiple journalistic organizations list the Michigan sightings as a top-ten news story for 1966.

And underneath all of this, in the part of Hynek that is neither the government consultant nor the public spokesman but simply the scientist who has been looking at this material honestly for eighteen years, something is happening that the press conference has not resolved and cannot resolve. He stood at that podium and said something he had been given to say, not something he had concluded. He knows the difference. He has always known the difference. The gap that has been opening between his private assessments and his public statements for nearly two decades has now, for the first time, become visible to the outside world — not because he has told anyone about it, but because the inadequacy of the explanation is so obvious to so many people that the gap has become the story.

On April 5, 1966, Hynek appears before the House Armed Services Committee alongside Air Force Secretary Harold Brown and Major Quintanilla. The hearing lasts one hour and twenty minutes. It is not the thorough congressional investigation that Ford called for — Ford himself later notes he would have preferred full hearings with witnesses — but it is a congressional hearing, and Hynek is in the room, and there is a statement in front of him that he has prepared himself. He reads it. It is, the record notes, certainly not dictated by the Air Force. In it, Hynek calls for the formation of a civilian panel of physical and social scientists to investigate UFOs seriously, for the express purpose of determining whether a major problem actually exists. He has not seen evidence to confirm extraterrestrial visitors, he says carefully. But he says, equally carefully, that the subject merits genuine scientific attention. He is not recanting swamp gas. He is doing something more consequential: he is, for the first time in a government context, publicly contradicting the institutional position that there is nothing here worth investigating.

The Air Force, under congressional pressure, responds by announcing it will commission an independent scientific study. By October the University of Colorado has the contract, and Edward Condon, a distinguished physicist, is named to lead what will become known as the Condon Committee. Hynek greets this news with something like relief. An independent scientific study is what he has been privately arguing for years should happen. He writes about it publicly with what one observer later notes is almost embarrassing enthusiasm — as though, having spent eighteen years watching the institutional process fail, he has convinced himself that this time the process will work. He is, in this moment, still enough of an institutionalist to believe in the redemption of institutions.

In December 1966, the Saturday Evening Post publishes an article by Hynek titled Are Flying Saucers Real? It is the first time he has addressed the subject at length in a mainstream publication under his own name with his own conclusions rather than the Air Force's. He is careful — he is always careful — but the direction is unmistakable. He argues that a residue of UFO reports deserves serious scientific attention. He challenges the Robertson Panel's methodology. He does not say flying saucers are real in the sense that would make headlines. He says the question is real, which is a different and, in some ways, braver claim, because it cannot be dismissed as crankery and cannot be accommodated by the institutional machinery that has been managing this subject for eighteen years. The question is real means the investigation has been inadequate. The investigation has been inadequate means the institution has failed. He is saying this in the Saturday Evening Post under his own name and his own credentials, and he is not taking it back.

Two words follow him everywhere now. Swamp gas. They appear in introductions at conferences, in the first paragraphs of newspaper profiles, in the punchlines of comedians who do not know his name but know the phrase. He will carry them for the rest of his life as the emblem of the public performance he delivered in a moment of institutional pressure — the thing he said that he did not believe, that everyone knew he did not believe, that became the public record of a private capitulation he has now, quietly but irreversibly, begun to repudiate.

In 1984, he will tell Omni magazine that the Michigan incident was a turning point. He will not elaborate extensively. He does not need to. The turning is visible in the record — in the congressional testimony, in the Saturday Evening Post article, in the steady accumulation of public statements through 1967 and 1968 that diverge increasingly from the institutional line. The man who made one phone call to Washington and came back with swamp gas is still there, in the historical record, permanent and undeniable. But he is no longer the man Hynek is choosing to be.

The Condon Committee begins its work. Hynek watches it carefully, with the specific attention of a man who knows what good investigation looks like and is waiting to see whether this will be it.

Chapter Seven

The Condon Committee

Hynek greets the announcement of the University of Colorado study with what one observer later describes as almost embarrassing enthusiasm. This is, in retrospect, a useful detail — not because it embarrasses him, but because it reveals the depth of what he has been hoping for. He has spent twenty years inside a broken process. He has watched institutional machinery turn evidence into non-evidence with a regularity that has ceased to surprise him. And now a respected physicist at a major university, funded by the Air Force but formally independent of it, is going to examine the Blue Book data without the constraints that have corrupted every previous examination. Hynek writes about the Condon Committee in his 1966 Saturday Evening Post article in terms that suggest genuine relief. He characterizes Condon's perspective toward UFOs as basically negative, but assumes, with what he will later recognize as misplaced faith, that Condon's opinions will change once he familiarizes himself with the evidence in some of the more puzzling cases. He has watched evidence change minds before. He has watched it change his own.

He is not wrong about Condon's negativity. He is wrong about its revisability.

Robert Low, the assistant dean who serves as the committee's project coordinator, writes a memo on August 9, 1966 — weeks before the University of Colorado formally agrees to undertake the study, before the contract is signed, before a single case file has been opened. He sends it to two University of Colorado administrators who need persuading that the project is worth the university's involvement. The memo's purpose is reassurance: Low is telling the administrators that the study will be safe, that the university's reputation will not be compromised, that the outcome is manageable. The trick, he writes, will be to describe the project so that, to the public, it would appear a totally objective study but, to the scientific community, would present the image of a group of non-believers trying their best to be objective but having an almost zero expectation of finding a saucer. The study would be conducted almost entirely by non-believers who, though they could not possibly prove a negative result, could and probably would add an impressive body of thick evidence that there is no reality to the observations.

Low files the memo. The university signs the contract in October 1966. The study begins. The memo sits in the project's open files, unnoticed, for almost a year.

Hynek is invited to Boulder to brief the committee on his own views about UFOs and does so. He brings the weight of twenty years and thousands of cases and the carefully calibrated language of a man who has learned, at considerable cost, to say exactly what the evidence warrants and nothing more. He leaves Boulder cautiously hopeful. The committee members seem intelligent. Some of them seem genuinely open. Condon himself is largely absent from the project's daily operations — his other responsibilities leave Low making most of the decisions — but the work is proceeding and the cases are being examined and there is at least the form of scientific investigation taking place. Hynek watches from Evanston and waits.

What he sees, as 1967 progresses, begins to concern him. Condon makes statements at a Sigma Xi lecture in January 1967 that strongly indicate his conclusions are already in place — he tells the audience that UFOs are not the business of the Air Force and implies the whole matter is beneath serious attention, this at a time when his committee has been examining case files for three months. Two of the committee's own investigators, David Saunders and Norman Levine, are growing increasingly troubled by what they observe from inside the project. On December 12, 1967, Hynek and the atmospheric physicist James McDonald — who has become the most scientifically rigorous and publicly outspoken critic of Blue Book and the most formidable advocate for serious investigation — meet informally in Denver with Saunders and Levine. They discuss, at that December meeting, the possibility of forming an independent organization to continue serious UFO investigation regardless of what the Colorado study concludes. The institution cannot be trusted. They have all reached this conclusion independently. They are now saying it to each other out loud.

Shortly after that December meeting, McDonald learns of the Low memo. He is shocked — his own word — and writes to Low expressing his concerns about the negative bias it reveals. Low hangs up the phone when McDonald calls him. Saunders and Levine, identified as the source of the memo's disclosure, are fired from the committee in February 1968. In May, Look magazine publishes a full account of the internal chaos under the headline Flying Saucer Fiasco, describing the project as a five-hundred-thousand-dollar trick. NICAP severs its relationship with the committee the same month. Low resigns. The project, which was supposed to be the independent scientific investigation that would settle the matter once and for all, has become a public scandal about the integrity of the investigation itself.

Hynek writes about the Low memo with what, under the circumstances, is considerable restraint. He argues that Low wanted his university to get the contract and needed to convince the administration to take it, which is accurate as far as it goes. He does not characterize Low as a plotter or conspirator. What he does say, with the precision of a man who has been watching institutional processes corrupt scientific inquiry for twenty years, is that the Condon Report settled nothing. And then, reading as carefully as he has learned to read, he says something that only a scientist who has spent twenty years inside the Blue Book system could say: carefully read, the report constitutes about as good an argument for the study of the UFO phenomenon as could have been made in a short time. This is not sarcasm. It is a rigorous observation. The body of the Condon Report — the case investigations conducted by Saunders and Craig and the others before the internal collapse — documents dozens of cases that resist conventional explanation. The report's conclusions, written by Condon and placed at the front of a 1,485-page document that most readers will not read past the summary, say one thing. The evidence inside the report, if anyone examines it, says another.

The Condon Report is formally released in November 1968 and made public in January 1969. The National Academy of Sciences endorses its conclusions. The Air Force, which has been waiting for exactly this endorsement, moves quickly. On December 17, 1969, the Secretary of the Air Force announces the termination of Project Blue Book. From 1947 to 1969, 12,618 sightings were reported to Blue Book. Of these, 701 remain in the files labeled Unidentified — cases that survived every conventional explanation the project's analysts could construct, cases that Hynek knows better than anyone because he processed most of them himself. They will sit in the National Archives, unexamined by official science, for the foreseeable future.

Three weeks before the termination announcement, a memo dated October 20, 1969, is signed by Brigadier General Carroll Bolender, the Air Force's Deputy Director of Development. The memo recommends closing Blue Book on the grounds that it cannot be justified on national security or scientific grounds. It also contains, in its closing paragraphs, a sentence that will not become public for nearly a decade — released only through a Freedom of Information Act request in 1979. Reports of unidentified flying objects which could affect national security, the Bolender memo states, are not part of the Blue Book system. They are handled through standard Air Force procedures designed for this purpose. Bolender does not explain what these procedures are. He does not identify the agencies involved. The sixteen pages attached to his memo that would presumably have shed light on this are missing from the Air Force files.

What the Bolender memo reveals, for anyone reading it carefully, is that Blue Book was never the real investigation. The cases that most warranted investigation — the cases that could affect national security, which is to say the most credible and significant cases — were not going through Blue Book at all. They were going somewhere else, through channels that remain unidentified. Project Blue Book, in which Hynek invested twenty-one years of his professional life, was the public face of a system designed to produce reassuring conclusions about a selection of cases from which the most important material had already been removed.

Hynek does not know about the Bolender memo in December 1969. He will not know about it for years. What he knows in December 1969 is that Blue Book is closed, that twenty-one years of his consulting work have produced an official conclusion that he does not believe is warranted by the evidence, that the independent study that was supposed to settle the matter was corrupted before it began, and that 701 cases the Air Force itself classified as Unidentified are now sitting in a government archive where no one is going to look at them.

He publishes his review of the Condon Report in the Bulletin of the Atomic Scientists. It is careful, substantive, critical in the precise way that a scientist is critical — not angry, not dismissive, but methodologically specific about where the study's methodology failed and what its actual findings, as distinct from its stated conclusions, demonstrate. He has been saying versions of this privately for years. He is saying it now in print, in a peer-reviewed publication, under his name and credentials, without the Air Force contract that once required him to hedge.

He also, sometime in this period, says something to Omni magazine that is the most honest summary of the Blue Book years he will ever produce. He says: I was a thorough skeptic, and I'm afraid I helped to engender the idea that it must be nonsense, therefore it is nonsense. The date of the interview is 1985, a year before his death, but the reckoning it represents has been building since at least 1966. He is not absolving himself. He is not dramatizing his guilt. He is doing what he has always done, which is saying exactly what the evidence — in this case the evidence of his own life — actually shows.

There is one institutional commitment remaining. The December 1967 Denver meeting where Hynek and McDonald and Saunders and Levine discussed forming an independent organization has never been forgotten. The Condon Committee's failure has, if anything, sharpened the need. Serious investigation of the phenomenon cannot happen inside government institutions that have decided in advance what the investigation will find. It cannot happen in the UFO enthusiast subculture that Hynek has spent two decades carefully distinguishing himself from. If it is going to happen at all it will require a new kind of institution: scientifically rigorous, methodologically accountable, independent of the Air Force and the enthusiast community alike, run by people with credentials serious enough that the scientific establishment cannot simply wave them away.

Hynek is fifty-nine years old. He has given twenty-one years to an institution that used him and discarded the results of his honest work. He has watched two independent investigations — the Robertson Panel and the Condon Committee — reach predetermined conclusions. He has 701 unidentified cases in the files that no one official will ever examine again. He has the invisible college, and McDonald, and Vallée, and a December meeting in Denver where four scientists agreed that whatever happened next would have to happen outside official channels.

He is done being the company man. He has a great deal of work left to do.

Chapter Eight

Going Public

On July 29, 1968, J. Allen Hynek walks into the Rayburn House Office Building in Washington and takes his seat before the House Committee on Science and Astronautics. He is introduced by his congressman, Donald Rumsfeld of Illinois, who notes his distinguished career with appropriate brevity. Hynek has testified before congressional committees before — the April 1966 Armed Services hearing after swamp gas, where he called quietly for independent scientific investigation. This time is different. The Condon Committee is still nominally running, though its internal collapse is already public knowledge. The Air Force's consulting relationship with Hynek is technically still in place though it is effectively over. He has been invited by Congressman J. Edward Roush of Indiana, who considers UFOs a legitimate scientific question and has organized a symposium rather than the adversarial hearing that some had sought. He is speaking, Hynek tells the committee at the outset, as a private citizen and scientist and not as a representative of the Air Force.

That sentence is not a formality. It is a declaration. For twenty years, in every government context, the Air Force has been the frame through which Hynek's expertise flowed. His analyses were Air Force analyses. His conclusions were Air Force conclusions, or close enough that the distinction was a private matter. Now he is drawing the distinction publicly, in a congressional hearing, before the cameras and the record. He is his own scientist. He has his own views. They are not the Air Force's views, and he is done pretending otherwise.

Six scientists testify at the Roush Symposium. Hynek goes first, McDonald second, Sagan third, followed that afternoon by Hall, Harder, and Baker. Five of the six are of the opinion that there is a genuine scientific anomaly that warrants serious investigation. Only Sagan, while conceding unexplained cases, is more cautious about conclusions. McDonald's prepared paper — thirty-five years of case analyses condensed into a presentation that is the most rigorously documented argument for the phenomenon's reality that any professional scientist has yet put on the congressional record — draws sustained attention from the committee members. Hynek's testimony is quieter and more structural. He does not claim that UFOs are extraterrestrial. He does not claim to know what they are. He argues, with the methodological precision that distinguishes him from the enthusiast community, that the question itself is real and that the official investigation has been systematically inadequate to address it. He tells a member who asks about his corps of interested scientists that he has private information from a very large number of scientists who are interested. The invisible college, acknowledged on the congressional record.

The hearing is significant but not transformative. Roush loses his congressional seat in the next election and the UFO hearings he envisioned do not materialize. The symposium's papers are eventually published, edited by Sagan and Thornton Page, and become an academic reference work representing the range of scientific opinion as of 1968. But the larger scientific establishment does not follow. The repercussions for the AAAS and the broader scientific community are, in the assessment of the historian who studies the proceedings most carefully, nil. Hynek has spoken honestly to power. Power has listened politely and moved on.

Then, on December 19, 1969 — one week before the AAAS annual meeting in Boston where Hynek is scheduled to present — the Air Force announces the termination of Project Blue Book. The timing is not accidental. The announcement drops into the scientific community the week that community is gathering to hear arguments for taking UFOs seriously, flooding the discussion with the official conclusion that the matter is settled and no further investigation is warranted. It is a final institutional act of the kind Hynek has watched the Air Force perform for twenty-one years — using the timing of information release to shape the conclusion before the evidence can be examined.

He goes to Boston anyway. The 134th meeting of the American Association for the Advancement of Science convenes December 26 through 27, 1969, in a heavy Boston snowstorm that causes McDonald to miss the first day. Hynek presents on December 27. His title is Twenty-One Years of UFO Reports. He is standing before the most significant scientific audience he has ever addressed on this subject — peers, colleagues, people whose professional respect he has spent a career earning and whose contempt for the subject he has spent a career managing. He is sixty years old and he is done managing it.

He tells them that the intellectual climate today is enormously better for taking a good look at the phenomenon than it was even a few years ago. He tells them that the symposium itself is an example — that it would have been impossible to hold it a year or two ago. He tells them that had he, earlier, attempted to call for a major investigation, he would have lost credibility and undoubtedly all possible future effectiveness. He is being characteristically precise. He is explaining why he waited as long as he did, which is the only real answer he has to the question of why he maintained the institutional position for twenty-one years. He waited because the moment had not yet arrived when the credibility he had carefully preserved through those years could be spent on the thing it was being preserved for. He is telling his audience, with the understated honesty of a scientist summarizing a twenty-one-year experiment, that the moment has now arrived.

He also, at the AAAS symposium, does something that has never been done before in a formal scientific context. He presents a classification system. Six categories, arranged by type of observation and proximity to the witness, designed to bring the same taxonomic rigor to UFO reports that astronomy brings to stellar phenomena. The three distant categories — Nocturnal Lights, Daylight Discs, Radar-Visual — cover the large majority of sightings. The three close encounter categories are where the most scientifically significant data resides. Close Encounters of the First Kind: a visual sighting within approximately five hundred feet, sufficient proximity to eliminate most conventional misidentification. Close Encounters of the Second Kind: a sighting accompanied by physical effects — ground impressions, scorched vegetation, electromagnetic interference with vehicles or electronics, physiological effects on witnesses. Close Encounters of the Third Kind: the presence of animated entities in association with a UFO. Hynek deliberately chooses the phrase animated entities rather than aliens or extraterrestrials, because he has understood something important —

that the taxonomy's scientific value depends on its neutrality with respect to the ultimate explanation. He is classifying observations, not endorsing conclusions.

He also, wisely, declines at the AAAS symposium to discuss the CE3 category in any detail. The observations are real enough, the cases documented enough, but he knows that an audience of skeptical scientists who have followed him this far on the strength of radar-visuals and physical trace evidence will not follow him into territory involving humanoid occupants in the same session. He is a teacher as well as a scientist and he understands the difference between the argument that is true and the argument that can be heard. He plants the taxonomy and leaves the CE3 category for The UFO Experience, which he is already writing.

The results of the AAAS symposium are, by the most honest external assessment, limited. There are no mass resignations from the association in protest, no congressional demands for a new investigation triggered by the proceedings, no wave of scientific papers on UFOs appearing in the months that follow. The UFO situation that existed before the symposium remains much the same afterward. The scientific establishment has been given its opportunity to hear the best available scientific case for taking the phenomenon seriously. It has listened and not significantly changed its mind. This is not surprising. Paradigm shifts do not happen at symposia. They happen in the accumulation of published work, in the slow revision of the questions that younger scientists consider worth asking, in the gradual displacement of old frameworks by new data that the old frameworks cannot accommodate. Hynek understands this. He is not expecting the

AAAS symposium to settle anything. He is beginning a public record that did not previously exist.

The UFO Experience is published by Henry Regnery in Chicago in 1972. It is 276 pages plus plates, dedicated to the witnesses — an acknowledgment that the foundation of the entire enterprise is the human testimony that every official investigation has systematically undervalued. Hynek's fellow Caltech astronomer Bruce Murray reviews it in Science, the flagship journal of the AAAS, under the title Reopening the Question. The review is more respectful than dismissive, which is itself a measure of how far Hynek's public positioning has moved the institutional needle. Murray does not endorse the book's conclusions. He cannot ignore its existence.

The book does what its title promises. It treats UFO sightings as experiences — data, in the scientific sense, that requires analysis rather than dismissal — and applies to that data the full weight of Hynek's twenty-one years of professional assessment. The classification system developed at the AAAS symposium is here given its complete exposition. The distant categories establish the methodological baseline: what distinguishes a genuine anomaly from the large majority of sightings that are, as Hynek has always honestly said, conventional objects misidentified by untrained observers. The close encounter categories — now fully elaborated, including the CE3 material he held back at the AAAS — represent the core scientific argument. These are cases in which proximity eliminates misidentification, in which physical evidence exists independently of witness testimony, in which the consistency of description across unconnected observers constitutes a pattern that demands explanation.

He is careful, throughout The UFO Experience, about what he does and does not claim. He does not say UFOs are extraterrestrial. He does not say he knows what they are. He says that the residuals — the cases that survive rigorous application of every conventional explanation — constitute a genuine scientific problem that has been inadequately investigated. He says that the official investigation was, in his own expert judgment, totally inadequate to the task. He says that the scientific community owes it to the phenomenon's witnesses, who are often credible people reporting real experiences, to examine the data with the care it deserves. These are positions he has held privately for fifteen years. He is now saying them in print, under his name, in a book published by a reputable press and reviewed in the most significant scientific journal in the country.

The professional cost is real but not catastrophic. Some colleagues distance themselves. The giggle factor — the involuntary dismissiveness that has attached to any serious discussion of UFOs since the saucer craze of 1947 — is still present in his professional environment and he navigates it as best he can. He is not fired from Northwestern. He is not stripped of his credentials. He is, in the bluntest assessment, a tenured full professor at a major university with a real scientific reputation and enough institutional security to say what he actually thinks without losing everything. This is precisely why he spent twenty-one years building that security before spending it. The credibility he preserved through the Blue Book years, the credibility that Quintanilla called a liability because it made his debunking insufficiently reliable, is now the asset that allows his public position to be heard rather than dismissed.

The taxonomy enters the culture. The phrase close encounter of the third kind, which Hynek invented as a precise methodological category, begins its journey into common usage. It will travel farther and faster than he can predict. In 1973 a young director named Steven Spielberg is developing a film about UFOs. He contacts Hynek. He tells him the film is going to be called Watch the Skies. He changes his mind and adopts Hynek's phrase instead. Hynek is paid one thousand dollars for the use of the title, one thousand dollars for the rights to use stories from The UFO Experience, and fifteen hundred dollars for three days of technical consulting. He also has a brief cameo near the film's climax — a white-bearded scientist, pipe in hand, watching with undisguised wonder as the alien craft comes into view. He is playing himself. The role requires no preparation.

None of this — the AAAS symposium, the book, the Spielberg connection — resolves the question that has been accumulating in his private file for twenty-five years. The taxonomy is not an answer. It is a better set of questions. He has organized the data rigorously enough that the residuals can be discussed with scientific precision rather than dismissed with institutional convenience. That is a genuine contribution, probably the most durable thing he will produce, the thing that will still be in use fifty years later when congressional committees are holding hearings about UAP and government agencies are admitting that some percentage of reports cannot be explained.

But he is not finished. He has an institution to build.

Chapter Nine

How He Actually Investigated

In the April 1953 issue of the Journal of the Optical Society of America, while still employed as the Air Force's scientific consultant and still producing the analyses that kept Blue Book functioning as a debunking apparatus, J. Allen Hynek published an article titled Unusual Aerial Phenomena. Its institutional awkwardness was considerable. He was, simultaneously, the man the Air Force called to explain UFOs away and the author of what would become his best-known statement: Ridicule is not part of the scientific method, and people should not be taught that it is. The steady flow of reports, often made in concert by reliable observers, raises questions of scientific obligation and responsibility. Is there any residue that is worthy of scientific attention? Or, if there isn't, does not an obligation exist to say so to the public — not in words of open ridicule but seriously, to keep faith with the trust the public places in science and scientists?

This is, among other things, a methodological statement. It is Hynek asserting, in print, in a peer-reviewed journal, that the question of whether a scientific residual exists must be answered by the methods of science rather than the posture of dismissal. The article did not transform anything in 1953 — the institutional machinery ground on, the Robertson Panel met and concluded, the cases were processed and filed and forgotten at the same rate as before. But the sentence is there in the record, five years into his Blue Book tenure, eight years before the phenomenon will have fully seized his private attention, a quiet marker of the methodological conviction that was forming underneath the professional accommodation. He believed, even then, that the question deserved to be asked honestly. He was not yet in a position to ask it. He asked it anyway, in the journal where it would be least likely to cause disruption and most likely to be preserved.

This chapter is about how he actually asked it. The methodology that Hynek developed over twenty-one years of Blue Book consultancy and refined over fifteen subsequent years of independent investigation is the most durable contribution he made — more durable than any individual case solution, more durable even than the taxonomy, because it is the foundation on which the taxonomy rests. It is also, as noted in the outline of this book, the thing most books about Hynek skip. They cover the cases and the institutional drama and the cultural aftermath and treat the methodology as background. It is not background. It is the argument. It is what distinguishes Hynek from the enthusiasts and the cranks whose company he spent his career trying to avoid, and what gives his conclusions — tentative, qualified, epistemologically honest — the weight that theirs do not have.

The foundation is the witness. Hynek understood earlier than most investigators that the quality of witness testimony was the first and most critical variable in any UFO case, and that this quality could not be assessed from paperwork alone. In the Blue Book years, the majority of cases came to him as report summaries — forms filled out by Air Force personnel, condensed and abstracted from whatever the original witness had actually said. This abstraction was itself a form of analysis, and not always a reliable one, because the people writing the summaries had their own institutional incentives and their own assumptions about what a credible sighting looked like. Hynek learned to distrust summaries. He wanted the witness.

When he could get to a witness directly — and for the cases that mattered, he pushed hard for that access — he brought to the interview the same quality of attention he brought to any observational problem. His son Joel, who as a young ham-radio operator used to record many of his father's witness interviews, recalled later that what he got from his father above all was how important it was to keep an open mind. The recording habit itself is methodologically significant. Hynek wanted the witness's words on tape, not transcribed and filtered through his own summary, because he knew from twenty years of processing other people's summaries how much information the summarizing process destroyed. The specific words a witness chose, the hesitations, the corrections, the moments of uncertainty and the moments of absolute conviction — these were data that disappeared in transcription and had to be preserved if the assessment was going to be honest.

He developed, informally but consistently, a set of criteria for witness credibility assessment. Professional background mattered — not as a guarantee of accuracy, since trained observers can be wrong, but as a prior probability. A military pilot filing an official report about an unusual aerial object has several incentives not to do so: professional embarrassment, the knowledge that the report will be scrutinized by superiors, the reputational cost in a community that values rationality and precision. When a pilot files such a report anyway, the prior probability that it represents genuine experience rather than confabulation or misidentification is higher than it would be for a civilian witness with no such constraints. Hynek weighted this carefully. He also weighted the specificity of description — a witness who describes angular size, estimated distance, color gradations, and the precise sequence of observed motion is providing data that can be cross-checked against other observations and against the known characteristics of conventional phenomena. Vague descriptions told him little. Specific ones told him a great deal, even when what they told him was that he was dealing with something that the known characteristics of conventional phenomena could not accommodate.

Consistency was a separate criterion from specificity. A witness who tells the same story the same way across multiple interviews, without the elaboration and embellishment that tend to accumulate in repeated retellings of dramatic experiences, is demonstrating a kind of memorial reliability that a good investigator can recognize and should value. Hynek paid close attention to what a witness did not claim as well as what they did. The Lonnie Zamora case was exemplary in this regard — Zamora described what he saw, sequentially and precisely, and consistently refused to go beyond what he had directly observed. He did not speculate about origin. He did not embellish the appearance of the figures he had seen. He reported an experience with the clipped discipline of a law enforcement professional and stuck to that report through every subsequent interview. This quality of testimony — spare, specific, resistant to elaboration —

was for Hynek the strongest possible indicator that he was dealing with genuine observation rather than construction.

Astronomical cross-checking was the first analytical layer he applied to any sighting. As an astrophysicist, he could assess immediately whether a reported object's behavior and appearance were consistent with any known astronomical source — bright planets visible at the time, meteor trajectories, satellite passes, atmospheric phenomena associated with specific meteorological conditions. This was, in the Blue Book years, most of what his role required of him. The majority of reports yielded to this analysis because the majority of reports were, as he always honestly said, misidentifications of exactly these phenomena. Venus is extraordinarily bright under certain atmospheric conditions and has been responsible for more UFO reports than any other single celestial object. Meteors can be dramatically slow and colorful in ways that contradict most people's expectations. Satellite re-entries produce spectacular light shows that have no good natural analogy in most observers' experience. Hynek's astronomical expertise allowed him to dispose of these cases quickly and honestly, which is precisely why his failure to dispose of the residuals carries the weight it does. He had applied the tools most likely to find conventional explanations. The residuals survived those tools.

Atmospheric optics was a second analytical layer that he understood well and applied carefully. The atmosphere does strange things to light — temperature inversions can produce mirages and distortions, ducting can carry radar signals in ways that produce false returns, unusual refraction conditions can make distant lights appear close and bright and mobile. The category of conventional explanation that the atmosphere makes available is substantial, and Hynek took it seriously. Swamp gas was itself, in principle, an atmospheric optics explanation — luminous methane in specific conditions of temperature and humidity. The problem with swamp gas as an explanation for the Michigan sightings was not that atmospheric optics phenomena are impossible but that the specific atmospheric optics phenomenon invoked did not match the specific characteristics of what was reported. This is the methodological failure he recognized even as he delivered the explanation: using a real phenomenon as a placeholder for a genuine investigation rather than as the conclusion of one.

Photographic analysis was the methodological domain where Hynek was most consistently frustrated, for reasons he documented carefully. Photographs, which should in principle be the most direct evidence available, turned out in practice to be among the least reliable. Hoaxes could be constructed with relatively modest photographic skill, and distinguishing a hoaxed photograph from a genuine one required access to the original negative and controlled analysis under conditions that were rarely available. Even photographs that survived technical analysis often showed nothing more interpretable than an object of unknown nature at distance and frequently out of focus — which confirmed that something had been photographed but said nothing about what it was. Hynek wrote in his 1967 Playboy article that a proper investigation would produce, within a year of its initiation, photographs, movies, spectrograms, plaster casts of indentations if a landing occurred, and detailed measurements of brightness, speeds, and direction. The fact that no such investigation had been conducted was itself, he argued, a scientific failure. What existed instead was a large body of photographs of varying quality, most of which had been examined under varying conditions of rigor, yielding a small number that were genuinely unaccountable and a large number that were either hoaxes or inconclusive.

The Jaroslaw case of January 1967 was the most instructive of his photographic failures. A Michigan teenager named Grant Jaroslaw and his brother Dan produced several Polaroid photographs of what appeared to be a hamburger-shaped object, and Hynek — denied access to the original prints by the family — offered a hedged assessment that he could not identify the object as a hoax. Newspaper headlines ran Expert Sees No Hoax in Boys' UFO Photos across the country. Nine years later, a confession letter from the brothers confirmed what a more cautious investigator might have declined to comment on without the original negatives. Hynek's premature hedged endorsement, documented by researcher Curt Collins in detail, was a methodological error he would not repeat in the same form. It confirmed his conviction that photographic evidence, however appealing, required the original material and controlled analysis conditions before any assessment was worth making. He added this to his methodology as a hard rule rather than a caution.

Radar interpretation was the methodological domain where Hynek was perhaps most rigorous, precisely because radar evidence seemed to offer an objectivity that eyewitness testimony could not. A radar return does not have a psychology. It does not confabulate or exaggerate or mistake Venus for a spacecraft. When a radar blip appears on a scope, something physical has returned a signal, and the question of what that something is becomes an engineering problem amenable to analysis. Hynek understood the limitations of this reasoning — radar can be fooled by atmospheric ducting, by certain types of precipitation, by electronic interference —

and applied those limitations carefully. What he found in the subset of cases with both radar confirmation and independent visual observation was the most scientifically significant material in the Blue Book files. A radar-visual case in which multiple independent observers see the same object simultaneously confirmed by multiple independent radar stations is not susceptible to the standard dismissals. The witnesses might be unreliable. The radar cannot be unreliable in the same way at the same time across multiple stations. Hynek returned to radar-visual cases repeatedly in his published work because they represented the clearest available evidence that the residuals were not a function of human perceptual error.

Physical trace evidence was the final and in some ways most compelling methodological category, because it provided data that existed independently of any witness at all. Compressed and dehydrated vegetation at reported landing sites. Circular ground impressions consistent with the pressure distribution of a heavy object on specific landing gear configurations. Scorched soil samples with chemical compositions that laboratory analysis found unusual. Hynek documented these carefully, distinguished genuine trace cases from those where the chain of custody of evidence was too weak to support conclusions, and treated the verified cases as among the strongest available arguments that the phenomenon had physical reality. The CUFOS archives he built include a substantial collection of physical trace case documentation that represents some of the most methodologically sound material in the entire history of UFO investigation.

Electromagnetic effects — the interference with vehicle engines and electrical systems reported in close encounter cases — constituted a separate methodological category that Hynek found particularly significant. An automobile engine that ceases to function as a reported object passes within a certain distance and resumes normal function after the object departs is providing physical evidence of an interaction that is not dependent on the witness's interpretive faculties. The witness cannot unconsciously fabricate the engine stoppage. If the effect is genuine — and Hynek's methodology required him to distinguish genuine EM effects from engines that simply happened to fail at a dramatically convenient moment — it constitutes evidence of a physical field of a kind not produced by any known conventional phenomenon within the relevant distances. Hynek collected EM effect cases carefully, weeding out the cases where the engine failure could be attributed to the vehicle's age or prior mechanical state, and building a subset of cases where the timing, distance, and multiple-witness correlation made conventional explanation genuinely strained.

The residuals concept, which he first articulated publicly in the 1953 JOSA article and developed across all his subsequent work, was not simply a name for the cases that survived conventional explanation. It was a methodological claim about what those cases represented. In any large dataset, a residual that is randomly distributed — that has no systematic characteristics, that does not correlate with witness type or geography or time period or observational conditions — is plausibly explained as noise: the irreducible tail of cases where the conventional explanation exists but cannot be found with available information. What Hynek found, and what Vallée confirmed through computerized analysis in the 1960s, was that the UFO residuals were not randomly distributed. They had systematic characteristics. They clustered in ways that noise does not cluster. They correlated with witness credibility rather than against it — the best-documented cases from the most reliable observers were more likely to be in the residuals, not less. This is the opposite of what a noise explanation predicts. Noise cases should be the poorly documented ones, the vague reports from uncredentialed witnesses that resist analysis because there is not enough information to analyze. The UFO residuals were the well-documented cases from credentialed witnesses that resisted analysis despite abundant information. That is not noise. That is signal.

He was, throughout all of this, epistemologically honest in a way that the subject's history makes remarkable. He did not claim more than the evidence supported. His private assessments were more confident than his published ones, because he understood that published claims carry obligations that private ones do not, and that the obligation to the evidence required him to hedge where the evidence was incomplete even when his judgment was not. He wrote in 1967 that in all the years he had been looking upward he had never seen what he would so dearly love to see — that he would never report a UFO even if he did see one, not without a witness. This is not the statement of a credulous man. It is the statement of a methodologist who has learned, through twenty years of watching other people's credulous statements be dismantled by skeptics, that an investigator without corroborating evidence is not an investigator at all. He set the same evidential bar for himself that he set for every witness he interviewed. He never lowered it.

His overall epistemology, developed through the Blue Book years and sharpened in independent practice, can be summarized in terms that any working scientist would recognize. Start with the null hypothesis — assume conventional explanation until the evidence specifically excludes it.

Apply the available analytical tools systematically and honestly, including the ones most likely to find conventional explanations. Document the residuals that survive this process with the same rigor applied to the confirmations. Assess whether the residuals are randomly distributed or systematic. If systematic, investigate the pattern. Do not claim more than the pattern supports. And do not allow institutional pressure — whether from the Air Force, from the scientific establishment, or from the enthusiast community that would gladly accept overclaiming in the direction it preferred — to substitute for this process at any step.

It is a simple methodology. It is the methodology of science. The remarkable thing is not that Hynek applied it to UFOs. The remarkable thing is that for twenty-one years, inside the institution that employed him, he was largely not allowed to.

Chapter Ten

The Expansion of the Possible

Jacques Vallée is born on September 24, 1939, in Pontoise, in the Val-d'Oise north of Paris, a year before the German occupation of France and twenty-five years before he will watch Lonnie Zamora's landing site with J. Allen Hynek on a bright New Mexico afternoon and begin to understand that the two of them are going to take each other somewhere neither has gone before. He studies mathematics at the Sorbonne and astrophysics at Lille, and he begins his professional life as an astronomer at the Paris Observatory — the same institution where Le Verrier computed the orbit of Neptune from perturbations in Uranus and where the history of predictive science has a kind of sacred weight. He is a precise and methodical young astronomer with a gift for pattern recognition and an unusually broad intellectual range, and he is paying attention to things his colleagues would prefer he ignore.

At the Paris Observatory he watches, on one specific occasion, a superior erase tracking data. The data in question — magnetic tapes recording the trajectories of objects tracked by the observatory's instruments — contains returns that do not correspond to any known satellite or aircraft. The superior's reasoning, to the extent that it is articulated at all, is institutional: these anomalous returns are an embarrassment, a potential source of unwanted attention, a problem that disappears more cleanly than can be explained. The tapes are erased. Vallée keeps the memory. It lodges in him the way that specific acts of institutional suppression lodge in people who have been trained to believe that data is sacred — not as a grand conspiracy theory but as a precise, personal, unignorable example of what happens when professional incentives are allowed to override the obligation to preserve evidence. He does not become a crusader. He becomes more careful.

In 1962 he moves to the University of Texas at Austin, where he co-develops the first computerized map of Mars for NASA — work that establishes him as someone capable of large-scale systematic data analysis and that gives him both the credentials and the computational tools, he will need for the project already forming in his mind. In 1963 he arrives at Northwestern University, drawn there at least in part by the presence of J. Allen Hynek, and takes up a position as a systems analyst while pursuing doctoral research in what he will eventually call industrial engineering and computer science, though the work is really something closer to early artificial intelligence. He earns the PhD in 1967. In the four years between his arrival at Northwestern and the completion of his doctorate he builds, in collaboration with Hynek, the first interactive computerized database of UFO observations — a project that applies to the Blue Book case files the same systematic data analysis he brought to the Mars mapping work, and that produces, when the analysis runs, patterns that neither of them expected and both of them found impossible to dismiss.

What the database reveals is this: the distribution of sighting reports is not random. The cases do not cluster the way noise clusters — uniformly, without structure, in patterns that reflect only the distribution of observers. They cluster in ways that suggest periodicity, that suggest geographic concentration not always explainable by population density, that suggest a repeating pattern of characteristics across reports with no possible connection to each other. A phenomenon that generates random reports has the statistical signature of background noise. What Vallée finds is the statistical signature of a signal. He publishes his findings. The findings are not welcomed by the scientific community, which has spent the Robertson Panel era learning not to look at this material, but they are rigorous and they are documented and they do not go away.

Hynek follows this work closely because the quantitative confirmation of what he has been observing qualitatively for fifteen years is exactly what his own methodology has been building toward. The residuals he identified case by case, the patterns he noted across unconnected reports, the systematic characteristics that distinguished the significant cases from the noise — Vallée's analysis gives these observations a mathematical foundation. The invisible college now has a database. The anecdotes now have statistics. This is not nothing. This is the difference between a scientist saying I have noticed a pattern and a scientist saying I have measured a pattern, and the difference matters enormously in any argument with the scientific establishment that has spent twenty years dismissing the former.

Their intellectual relationship is close, productive, and conducted between two people who are similar enough in their rigorous habits to trust each other's analyses and different enough in their temperaments to push each other past their respective comfort zones. Hynek's comfort zone is the empirical: the case files, the physical evidence, the witness testimony, the statistical patterns. He is at home in the territory that his training has equipped him to inhabit, and he is reluctant to leave it without the kind of evidence that his training defines as sufficient. Vallée's comfort zone extends considerably further. He has read more widely in folklore, anthropology, and the history of anomalous experience than any astronomer has any particular reason to read, and what he finds in that reading is a pattern that the pure physical evidence approach cannot accommodate — the persistence, across cultures and centuries, of encounter experiences that share structural features with the modern UFO phenomenon. The little people of European folklore. The djinn of Islamic tradition. The demons and angels of Christian mystical literature. The visitors in the night that appear across human accounts of the uncanny from ancient Mesopotamia to twentieth-century New Mexico. Vallée is not arguing that these are the same phenomenon. He is arguing that the structural similarities are too consistent to be coincidental and too widespread to be a product of modern media influence, and that any theory of the UFO phenomenon that does not account for this historical continuity is almost certainly wrong.

He publishes Passport to Magonia in 1969 — the same year the AAAS symposium takes place, the same year Blue Book closes, the same year Hynek is finally free to say publicly what he has been thinking privately for a decade. The book is structured as a case for the continuity of the phenomenon across history and culture, drawing on a catalog of pre-twentieth-century encounter reports and demonstrating that the phenomenology of modern UFO encounters — the small figures, the paralysis, the time distortion, the physical traces, the obsessive quality of the memory afterward — appears in medieval accounts of fairy abductions, in Renaissance demonology, in the folklore of cultures that had no contact with each other and no possible source of narrative contamination from the modern saucer literature. It is, among other things, a methodological argument: if the phenomenon predates the cultural context that supposedly generates it, then the cultural context is not generating it. Something else is.

Hynek reads Passport to Magonia with a mixture of genuine intellectual excitement and professional caution that will characterize his relationship to Vallée's more radical propositions for the rest of his life. The historical data is real. The structural parallels are real. The argument that the extraterrestrial hypothesis — that UFOs are spacecraft piloted by beings from another planet, making the trip across interstellar distances for reasons they have so far declined to make plain — is too simple to account for the full phenomenology of the data, is, Hynek privately agrees, probably correct. The extraterrestrial hypothesis has always sat uneasily with him because it answers the question of what UFOs are without answering the question of why they behave the way they do, which is to say: strangely, inconsistently, more like a stimulus-response system designed to provoke reaction than like a survey mission designed to gather information. Vallée gives Hynek a framework for this unease. He does not give him certainty, and Hynek does not want certainty. He wants a better question, and Passport to Magonia provides one.

The control system theory, which Vallée develops through the late 1960s and early 1970s, is the most conceptually radical thing to come out of their collaboration. The idea, stated plainly, is that the UFO phenomenon functions as a control system for human belief and behavior — that its appearances are not random or exploratory but patterned in a way consistent with a program of graduated disclosure designed to influence the development of human consciousness and civilization. The phenomenon, in this view, is not simply visiting. It is managing. Managing what, toward what end, by what agency — these questions remain open. But the observable characteristics of the phenomenon, analyzed with the tools Vallée has been developing are more consistent with a feedback loop than with an exploration program. The phenomenon appears to be adjusting its presentation in response to human cultural development, showing us what we are prepared to see rather than what it actually is, calibrating its reality to the edge of our capacity to accept it without either believing it entirely or dismissing it entirely.

Hynek is interested in this and cautious about it simultaneously. Caution is methodological — the control system theory is, in the strict sense, unfalsifiable, which means it cannot be tested and therefore cannot meet the evidentiary standard he has spent his career maintaining. He can say the phenomenon is patterned without committing to a theory of why it is patterned. He can say the extraterrestrial hypothesis is insufficient without adopting the control system theory as its replacement. He holds the theory at arm's length, considers it seriously, discusses it candidly with Vallée in the conversations they continue throughout the 1970s, and declines to make it his own public position. This is not evasion. It is the same epistemological discipline that has governed his entire intellectual life: do not claim more than the evidence supports. The control system theory interests him as a hypothesis. He does not yet — and will not, quite, in his lifetime — have the evidence to claim it as a conclusion.

The Edge of Reality, published jointly in 1975 by Henry Regnery — the same press that published The UFO Experience three years earlier — is the most direct document of where their thinking has arrived together. The book is structured as a series of dialogues, mostly transcribed conversations between the two of them and, in some sections, the psychologist Arthur Hastings. The format is deliberate: they are showing their thinking in process rather than presenting finished conclusions, because the finished conclusions are not available and both of them are honest enough not to pretend otherwise. The opening declaration sets the stakes: the UFO represents an unknown but real phenomenon, and its implications take us to the very edge of what we consider the known and real in our physical environment. Perhaps UFOs signal the existence of a domain of nature as yet totally unexplored.

Within the dialogues, the intellectual divergence between them is visible in real time. Vallée pushes. Hynek follows up to a certain point and then pulls back toward the empirical ground. In one exchange that will be cited by everyone who subsequently writes about the interdimensional hypothesis, Vallée asks what other wild hypotheses they might make, and Hynek responds: there could be other universes with different quantum rules or vibration rates if you want. Our own space-time continuum could be a cross-section through a universe with many more dimensions. Think what a hard time you would have convincing an aborigine that right now, through this room, television pictures are passing. He is reaching, here, in a way that the younger Hynek — the empiricist, the positivist, the man who organized his professional life around the principle that knowledge comes from measurement — would have found professionally inadmissible. He is fifty-five years old and he is considering the possibility that the framework of physical reality that his entire scientific career has assumed is not the only framework, and that the phenomenon he has been studying for twenty-seven years might be native to the domain of that framework that his instruments cannot reach.

The Edge of Reality receives the reception that a book advancing these ideas in 1975 can expect: serious attention from the small community of people already engaged with the question, dismissal or silence from the scientific establishment, and a contemptuous review in the Skeptical Inquirer that characterizes their work as a house of cards built on hearsay evidence and careless scholarship. Hynek responds to this review in the pages of the same publication, noting with his characteristic precision that the reviewer has confused the category of reviewer who uses a review to air their own emotional bias with the category who engages with the main thrust of the work. The review does not change his mind. Nothing in the scientific establishment's response to his work changes his mind, because his mind has not been changed by responses — it has been changed by evidence, accumulated across twenty-seven years of careful looking, and the evidence has led him to a place that the scientific establishment has decided not to follow.

Vallée leaves Northwestern in 1969, moves to Stanford, and eventually to Silicon Valley, where he will apply his data analysis skills to the emerging computing industry with considerable success. He and Hynek do not stop talking. The Forbidden Science journals, which document Vallée's private thinking from the late 1950s through the mid-1970s, are full of Hynek — lunches, phone calls, late evenings working through cases, the specific quality of conversation between two people who respect each other enough to say what they actually think. What the journals reveal is that the gap between their public and private positions is considerably smaller than the public record suggests. Hynek in the journals is more willing to entertain Vallée's more radical propositions than Hynek in print. Vallée in the journals is more careful about the evidentiary basis for those propositions than Vallée in print. They are each other's best audience and best critic, which is the most useful intellectual relationship either of them has.

In 1978, they present together at the United Nations — part of a panel that includes NASA astronaut Gordon Cooper and a delegation of military and government officials — a formal proposal for international scientific cooperation on UFO investigation. The proposal goes nowhere. The UN is not ready to institutionalize UFO research any more than the Air Force was. But the fact of the presentation matters: two credentialed scientists, one American and one French, standing before an international body and asking, with the combined weight of their careers behind them, to be taken seriously. It is an act of institutional persistence that neither of them, in their earlier professional lives, could have imagined performing. Hynek is sixty-eight years old. He has been looking at this material, in one form or another, for thirty years.

What Vallée gives Hynek, in the final accounting, is permission. Not permission to believe anything specific — neither of them believe anything specific, in the sense of holding a fixed conclusion about what the phenomenon is. Permission to entertain, seriously and without professional embarrassment, the possibility that the phenomenon requires a framework that does not yet exist. Permission to follow the evidence past the point where the existing frameworks run out, into territory where a rigorous mind can proceed only by holding the question open and resisting the temptation of the premature answer. This is, in one sense, simply the methodological discipline that Hynek has always practiced. In another sense it is something more — a willingness to sit with genuine uncertainty at the edge of the possible, in the company of another careful mind, and not to flinch from what the evidence suggests even when what it suggests is strange beyond the range of any explanation currently available.

Hynek is not a mystic. He is not a philosopher in the technical sense. He is an astronomer who followed the evidence for thirty years and ended up at the edge of reality, looking out. Vallée is the person who stood beside him and said: yes. This is where the evidence leads. The question is what we do about it.

Chapter Eleven

Building the Institution

What Hynek wanted, when he imagined what a proper scientific organization for UFO investigation would look like, was essentially the anti-Blue Book. Not its opposite in conclusions — he had not decided what the conclusions should be — but its opposite in method. A place where the evidence drove the analysis rather than the analysis driving the conclusion. Where the case files were maintained with archival rigor rather than bureaucratic convenience. Where the investigators were credentialed professionals who brought their actual expertise to the cases rather than institutional representatives who brought the outcome their institution needed. Where the question was treated as genuinely open, which is the foundational condition of scientific inquiry and which Blue Book had never, in twenty-one years, actually managed to honor.

He had been thinking about this since at least December 1967, when he and McDonald and Saunders and Levine sat in Denver and agreed that whatever came next would have to come from outside official channels. He had been thinking about it through the Condon Committee's slow-motion failure, through Blue Book's closure, through the AAAS symposium and the publication of The UFO Experience and the growing public visibility that had made him the one scientist whom journalists called when they needed someone to explain why the question was legitimate. By May

1973 The UFO Experience had reached its fourth printing. New Scientist described him as the lead scientific voice making UFOs respectable. He was sixty-three years old and he had the credibility he had spent twenty-one years carefully accumulating and had never had a proper institution in which to spend it.

He tried Northwestern first. He approached the university administration about founding a center for UFO research under Northwestern's institutional umbrella, which would have given the project the credibility and the library access and the indirect affiliation with a major research university that he thought would help attract the serious scientists he wanted involved. The administration bristled. Whatever the scientific merits of the question — and the administration was not prepared to concede those merits were substantial — the reputational risk of official association with UFO research was not something Northwestern's leadership was willing to absorb. They declined. Hynek was not surprised. He had watched institutions make this calculation for twenty-one years.

He founded the Center for UFO Studies in 1973 without Northwestern, in Evanston, Illinois, in collaboration with Sherman Larsen — an insurance salesman who had been active in NICAP, the National Investigations Committee on Aerial Phenomena, and who brought to the partnership organizational energy and practical administrative competence that complemented Hynek's scientific authority. The combination was deliberate: Hynek was the scientific credibility, Larsen was the institutional infrastructure. The center was small, privately funded, and from the beginning structured around the principle that had driven Hynek's thinking since the invisible college — participation restricted to scientists and other professionals who donated their time and expertise. It was not a membership organization. It was not a saucer club. The ufologist James W. Moseley, who was not always generous in his assessments of the field's organizations, later wrote that CUFOS was the first real attempt to set up a private research group genuinely dedicated to scientific investigations and study of UFOs. That sentence, from someone who had seen every other attempt, is the most precise external assessment of what Hynek had built.

The organizational architecture reflected everything Hynek had learned from Blue Book's failures. Cases were screened before investigation rather than accepted wholesale. Investigators were assigned by area of expertise — astronomers to optical cases, radar specialists to EM interference cases, physicians to reports involving physiological effects. Physical trace evidence was handled with chain-of-custody protocols that Blue Book had never bothered to establish. Witness interviews were recorded rather than summarized. The UFOCAT database, which Vallée's computerized analysis had proved the value of, was maintained and expanded — it would eventually catalog over one hundred thousand global reports, becoming the world's largest repository of UFO sighting data and the most systematic tool available for the kind of pattern analysis that Hynek and Vallée had been doing informally for a decade. An academic publication, the Journal of UFO Studies, was established for peer-reviewed analysis. A public newsletter, the International UFO Reporter, began in 1976 as a way of maintaining connection with the broader community of interested people without compromising the organization's scientific standards.

In 1976 CUFOS hired its first full-time investigator, a young astronomer named Allan Hendry, who brought to the work the same methodological rigor that Hynek demanded of everyone associated with the center. Hendry's 1979 report on his ongoing research — a comprehensive analysis of hundreds of cases investigated under CUFOS protocols — was the most systematic study of UFO reports ever produced by a single investigator working within a defined methodology, and its conclusions were characteristically honest: ninety to ninety-five percent of reports resolved as conventional phenomena upon careful investigation; the remaining five to ten percent were genuine residuals that survived rigorous analysis. The report was not what the enthusiast community wanted — they wanted the percentage to be higher and the conclusions to be more dramatic — and it was not what the skeptical community wanted either, because a five-to-ten percent residual from a carefully filtered dataset of cases is not noise. It is a finding. Hendry presented it as a finding and did not apologize for what it showed or what it didn't show. Hynek considered this exactly right.

CUFOS's first important case arrived before the center had fully established itself, in October 1973, the same month as its founding — a wave of UFO reports concentrated in the American South that produced within a single week two cases that Hynek would later describe as the most notable he had encountered in a single seven-day period in his entire career.

On the evening of October 11, 1973, Charles Hickson, forty-two years old, a shipyard foreman in Pascagoula, Mississippi, and Calvin Parker, nineteen, a welder at the same yard, were fishing off the west bank of the Pascagoula River when they heard a buzzing sound and saw a glowing, oval-shaped craft hovering above the ground forty feet away, eight to ten feet high and perhaps thirty feet across, emitting a blue light. Three beings emerged — five feet tall, gray-skinned, with claw-like hands and no visible eyes — and the two men felt themselves paralyzed and lifted toward the craft. They were examined inside by what Hickson later described as a large mechanical eye and then returned to the riverbank. The craft departed. Hickson and Parker, unable to explain what had happened to them and uncertain whether they would be believed, drove first to the local newspaper office — finding it closed — and then to the Jackson County Sheriff's Department.

The sheriff listened to them, found them distressed and coherent, and made a practical decision: he put them in a room wired for sound and left them alone, expecting that if their account was fabricated, they would relax and admit it to each other when they thought no one was listening. The tape recorded instead two men in genuine distress, unable to make sense of what they had experienced, with no suggestion of hoax in anything they said. You sure you're not going crazy, Calvin, Hickson asked. No, Mr. Charlie, Parker replied. I know what I saw. I just want to wake up and not remember any of it. The sheriff played the tape and found he had no good category for what he was hearing.

The wire services picked up the story within days, and it became national news. James Harder of the Aerial Phenomena Research Organization arrived first and conducted hypnotic regression interviews with both men, having to terminate the session with Hickson when Hickson became too frightened to continue. Hynek arrived representing the newly established CUFOS and interviewed the men separately, applying the witness evaluation methodology he had developed over twenty-five years. Hickson passed a polygraph. Parker, emotionally shattered, eventually did as well. Both men, examined at Keesler Air Force Base, were cleared of radiation exposure. Hynek's assessment, delivered to the press at a joint appearance with Harder, was careful but clear: these men have had a very real, frightening experience. They are absolutely honest. They have had a fantastic experience. There was definitely something here that was not terrestrial. These are not the words of a converted believer — they are the words of a methodologist who has applied his criteria to a case and reported what the criteria found. Hynek could not tell them what had happened to Hickson and Parker on the Pascagoula River. He could tell them that the men were telling the truth about what they believed had happened, and that the truth they were telling did not fit any conventional category he knew.

Seven days later, on the night of October 18, four members of a United States Army Reserve helicopter crew flying from Columbus to Cleveland in a UH-1H Huey encountered something over Mansfield, Ohio, that will remain in the case files as among the most physically documented encounters in the history of American UFO investigation.

Captain Lawrence J. Coyne, thirty-six years old, with nineteen years of flying experience, was commanding the flight from the right front seat. At approximately eleven p.m., ten miles south of Mansfield, Sergeant Robert Yanacsek noticed a single red light on the horizon that appeared to be on a converging course with the helicopter. Within thirty seconds it was apparent the light was closing at a speed estimated at over six hundred knots. Coyne took the controls and initiated a powered descent from twenty-five hundred feet, simultaneously contacting Mansfield approach control, which acknowledged once and then went silent. The crew prepared for impact. The object stopped directly above and ahead of the helicopter — a gray, metallic, cigar-shaped craft approximately sixty feet long and twenty feet high, filling the entire windshield. A green beam swung from beneath it and swept through the cockpit. The whole cabin turned green, Coyne later said. The crew watched in silence. After several seconds the object accelerated to the west, made a forty-five degree turn, and disappeared toward Lake Erie.

What happened next is the case's most technically significant detail. Coyne had been maintaining a controlled descent with the collective — the helicopter's main lift control — in the full-down position. He had made no attempt to climb. When the object departed and he checked his instruments, the altimeter read thirty-five hundred feet. The helicopter had climbed almost two thousand feet while the controls were set for descent, without the crew experiencing the g-forces that a two-thousand-foot climb under power would produce. No conventional aerodynamic explanation has been offered for this. The helicopter's magnetic compass had begun rotating. The radio, which had failed during the encounter, was working normally again. Below them, unknown to the crew, a family driving on Route 430 — Emma Delong and her four children — had stopped their car and watched the event unfold, saw the green light paint the helicopter and the surrounding terrain, and later provided ground-witness confirmation that precisely matched the crew's account in timing, sequence, and described appearance.

Coyne reported the incident to a Cleveland journalist immediately upon landing, before any military authority had the opportunity to manage the disclosure. He appeared on the Dick Cavett Show and described what had happened with the clipped precision of an experienced military pilot who is accustomed to filing accurate incident reports and has no particular investment in making the incident more dramatic than it was. He described himself as a skeptic of UFOs. He had no explanation for what he had witnessed. CUFOS assigned Jennie Zeidman — a former Ohio State astronomy student who had briefly worked on Blue Book and knew the cases — to conduct the primary investigation. Her report, exhaustive and methodologically sound, confirmed every verifiable detail of the crew's account and the ground witness testimony, and concluded that no conventional explanation was available. CUFOS assessed the Coyne case as the most credible incident in the wave of 1973. The National Enquirer's Blue-Ribbon Panel of scientists awarded it the five-thousand-dollar prize for the most scientifically valuable UFO report of the year.

The 1973 wave was, as Hynek later said, a memorable year — containing within one week two classic cases that together demonstrated what CUFOS was built to do: bring rigorous methodology to cases that the official system had already decided it would not examine. Both cases have remained in literature as among the most carefully documented encounters of the decade. Both involved witnesses of high credibility with no investment in the phenomenon who reported experiences they could not account for and stuck to those accounts under sustained scrutiny. Both were investigated by CUFOS using standards that Blue Book had never applied. Both remain unexplained.

Six years later, in the early hours of August 27, 1979, Marshall County Deputy Sheriff Val Johnson was on patrol on a desolate stretch of highway in northwestern Minnesota, eleven miles west of Stephen, when he saw a bright light low to the ground in the distance. He thought it might be a downed aircraft. He turned toward it. The light rushed at him with a speed he could not estimate — not fast in the way that moving objects are fast, but instantaneous, as if distance had been removed rather than crossed. The last thing he heard was the sound of breaking glass. He regained consciousness thirty-nine minutes later, his car sideways across the road, and radioed dispatch: Something just hit my car.

What the responding officers found documented itself: a cracked windshield broken from outside in, a shattered headlight, a bent antenna, a damaged red emergency light. Skid marks eight hundred and fifty-four feet long. Johnson himself with eye injuries that his examining physician compared to welder's burns. And a detail that no investigator has ever satisfactorily explained: the clock in Johnson's patrol car and the watch on his wrist were both running fourteen minutes slow. Both had been synchronized with headquarters before the incident. They were both wrong by exactly the same amount.

CUFOS sent Allan Hendry to Minnesota. He examined the vehicle, conducted magnetic field tests, interviewed Johnson and the officers who responded to his call, and assembled a report that CUFOS considered significant enough to dedicate nearly the entire September-October 1979 issue of the International UFO Reporter to the case, delaying publication specifically to get the investigation complete. Honeywell engineers and a Ford Motor Company crash investigator examined the car. The Ford examiner said he had not seen anything like the damage pattern before and that it was consistent with an impact from outside by an object of sufficient force to stress but not shatter the glass — a description that matched no known conventional phenomenon in the area at the time. The FAA and the Air Force confirmed no aircraft were in the vicinity. Hendry's conclusion was his characteristic one: the biggest mystery about the Val Johnson case is trying to find one neat explanation for something that could behave the way he described yet create the kind of damage that we analyzed and discovered. No neat explanation has been found. Johnson's patrol car — the single most physically documented piece of UFO-related evidence in the CUFOS archive — is now on permanent display at the Marshall County Historical Society Museum in Warren, Minnesota.

These three cases — Pascagoula, Coyne, Val Johnson — represent what CUFOS was built to investigate: cases where the witnesses are credible, the physical evidence is real, the conventional explanations have been genuinely applied and have failed, and the residual is therefore something that honest science must acknowledge rather than file and forget. They also represent the range of the phenomenon as Hynek had come to understand it — a close encounter of the third kind involving humanoid entities, a close encounter of the first kind involving military personnel and physical effects on a vehicle and instruments, a close encounter of the second kind involving physical damage, medical effects, and an anomaly in time. All three occurred outside official channels. All three were investigated by CUFOS with methodological rigor that the Air Force had never applied. All three remain in the files as the kind of data that the question requires but that no official process has yet been willing to confront.

This is what building the institution meant in practice. Not conferences and position papers and battles for scientific legitimacy, though there were all of those. It meant showing up in Pascagoula and in Mansfield and in Marshall County, Minnesota, with the recording equipment and the methodology and the willingness to say honestly what the evidence showed, in the full knowledge that what the evidence showed was going to be called nonsense by most of the scientific establishment and not enough by most of the enthusiast community. Hynek occupied the difficult middle ground between those two positions for the last thirteen years of his life, and CUFOS was the institution he built to hold that ground when he was gone.

He had decided, at some point in the years between Blue Book's closure and CUFOS's founding, that the institution mattered more than his personal vindication. The cases were going to keep happening whether or not anyone with adequate methodology was available to investigate them. The data was going to keep accumulating whether or not anyone with adequate rigor was available to analyze it. He could not make the scientific establishment pay attention. He could make sure that when it eventually decided to pay attention, there would be something to pay attention to.

Chapter Twelve

The Famous Astronomer

In the fall of 1973, the same season that CUFOS was founded and Pascagoula and Coyne were happening, a young filmmaker named Steven Spielberg was developing a deal with Columbia Pictures for a science fiction film about UFOs. He had been thinking about making this film, in one form or another, since he was a teenager in Phoenix — since the night his father woke him and took him to a field outside the city to watch a meteor shower, and he lay on his back in the Arizona dark and felt the sky as something vast and alive and full of possibility. He had made an amateur 8mm UFO film at sixteen, titled Firelight, which he screened in a local theater for one night and turned a modest profit. He had never stopped thinking about the subject. He was now twenty-six years old, rich and powerful beyond any reasonable expectation after Jaws, and he was going to spend that power on the most personal film he could imagine.

He had been reading Hynek. He had absorbed The UFO Experience in the way that certain books are absorbed — not skimmed but inhabited, carried around in the mind, the argument tested against everything else one knows and found, repeatedly, to hold. What he found in Hynek was the thing he had been looking for: a serious scientist, a credentialed professional, someone whose judgment could not be dismissed by the people who wanted to dismiss the subject, saying that the evidence was real and the question was legitimate and the official investigation had been systematically inadequate. Hynek gave Spielberg a professional's point of view, as Spielberg later said, on field reporting. He gave the film intellectual ballast before a single frame was shot.

The film's title went through several incarnations. Spielberg's original working title was Watch the Skies, from a line of dialogue at the end of the 1951 science fiction film The Thing from Another World. An early draft by Paul Schrader was called Kingdom Come. Another draft, by John Hill, was called Meeting of the Minds. In early 1976, while principal photography was being prepared, the title reverted briefly to Watch the Skies — a trade paper item in February 1976 announced this, apparently without Spielberg's approval. What followed was a letter from Hynek's attorney. The letter raised the question of rights to the phrase Close Encounters of the Third Kind, which Hynek had coined in The UFO Experience and which Spielberg had been using in script drafts without any formal arrangement. A settlement resulted in Columbia Pictures purchasing the rights to The UFO Experience and hiring Hynek as technical consultant for the film. The title was restored. It would not change again.

Principal photography began May 16, 1976. The large-scale contact sequences were filmed not at Devil's Tower, Wyoming, where some of the exteriors were shot, but inside a vast aircraft hangar in Mobile, Alabama — the largest interior filming space in the world at the time, large enough to contain the mothership set without any visible edges to the frame. On July 23, 1976, J. Allen Hynek arrived on the Mobile set. Bob Balaban, who played the cartographer-interpreter David Laughlin in the film and who was keeping a diary that he would later publish, recorded his arrival:

Hynek, he wrote, was wearing a Hawaiian shirt and didn't look like a scientist except for his neatly cropped Van Dyck beard which made him look a little like the Wizard of Oz. He thought the movie would help the UFO cause since Spielberg had done such thorough research and based so much of the film on actual events. No photographs could be taken on the set — Columbia's security was strict — so Hynek sat quietly in his canvas chair aiming a small tape recorder in the direction of the filming. Since he couldn't take pictures he was taping the sounds in the hangar to help him remember this day.

That evening Hynek gave a lecture to the cast and crew — Richard Dreyfuss and Melinda Dillon were there, along with about forty others — about the phenomenon and its documented cases. He was, Balaban recorded, deeply earnest and the audience was rapt. For a sixty-six-year-old astronomer from Evanston, Illinois, who had spent two decades in the institutional shadows of Project Blue Book, to be standing in a Mobile airplane hangar lecturing to the star of Jaws about the scientific residuals of UFO investigation, in the service of a film that was going to be seen by hundreds of millions of people — this was not a situation that any version of his earlier life could have predicted. He appears to have found it genuinely delightful.

He had objections, which he voiced. He told Spielberg that associating the film's premise with the disappearance of the Navy TBM Avengers of Flight 19 in the Bermuda Triangle — the opening sequence of the film, in which the aircraft are found immaculately preserved in the Sonoran Desert — was not consistent with the phenomenon as documented. The Bermuda Triangle was, in Hynek's assessment, a media creation rather than a genuine anomaly, and conflating it with serious UFO evidence undermined the film's credibility with the audience he most wanted to reach. Spielberg listened and kept the sequence. He was making a film, not a documentary. But he kept Hynek, which tells you something about what Spielberg valued in the collaboration: not endorsement, not the authority of credentials weaponized to validate a predetermined premise, but the genuine thing — the instilled sense of professional seriousness that prevented the film from sliding into the credulous sensationalism that Hynek had spent his career trying to distinguish himself from.

The film premiered November 17, 1977, at the Ziegfeld Theater in New York. It grossed over three hundred million dollars worldwide and became the third highest-grossing film of 1977, behind only Star Wars and Smokey and the Bandit. It received multiple Academy Award nominations. In December 2007 the Library of Congress deemed it culturally, historically, and aesthetically significant and selected it for preservation in the National Film Registry. The French director François Truffaut, who plays the lead scientist Claude Lacombe in the film, was cast partly because he embodied the European intellectual seriousness that Spielberg wanted to associate with UFO investigation — and partly because Truffaut himself was fascinated by the subject. The character Lacombe was drawn, at least in part, from Vallée, who had become the model of the credentialed European scientist taking the phenomenon seriously from outside the American institutional framework that had so consistently suppressed it.

Hynek's cameo begins at two hours, two minutes, and fifty-seven seconds into the 1998 Collector's Edition — just after the pilot and crew of Flight 19 emerge from the landed mothership. He strolls to the front of the crowd of scientists and military personnel watching the contact sequence, brushes his goatee with his hand, and inserts his pipe into his mouth. He is credited on IMDB as Smoking Pipe at Landing Site. He is playing himself — a scientist, watching with undisguised wonder as the extraordinary becomes real — and the role requires nothing he does not already have. His expression in the seconds of screen time is not performed. It is the expression of a man who has spent thirty years believing that this moment, in some form, was possible.

Hynek's assessment of the film, offered to the press at the time of its release, was characteristic — careful, measured, specific about what it got right and what it did not. Even though the film is fiction, he said, it's based for the most part on the known facts of the UFO mystery, and it certainly catches the flavor of the phenomenon. Spielberg was under enormous pressure to make another blockbuster after Jaws, but he decided to make a UFO film. He put his career on the line. The phrase he put his career on the line is, from Hynek, a form of professional recognition. He understood what it meant to stake credibility on an unfashionable position. He had been doing it for a decade.

The film did something to the culture that Hynek's books and lectures and congressional testimony had not been able to do, and could not have done, because it operated in a different register entirely. A book argues. A film experiences. Close Encounters of the Third Kind asked its audience not to evaluate evidence but to feel possibility — to sit in a darkened theater and experience, for two hours and seventeen minutes, what it would be like to live in a world where the phenomenon was real and the contact was coming. Hundreds of millions of people had that experience. The phrase close encounter entered the language. The taxonomy Hynek had developed as a methodological tool, a classification system for organizing field reports, became the most famous contribution he had ever made, known to people who had never heard his name and would not have read his books under any circumstances.

The visibility had a complicated relationship with the scientific credibility he had been building. The giggle factor — the involuntary dismissiveness that attached to any serious discussion of UFOs — was not eliminated by Close Encounters. It was, if anything, reinforced in certain quarters, because the film's enormous popular success associated the phenomenon more firmly with entertainment and fantasy in the minds of the scientific establishment that Hynek most wanted to persuade. The people who came to his lectures after 1977 were different from the people who had come before — more numerous, more enthusiastic, and sometimes more interested in the wonder of the film than in the rigorous methodology that was Hynek's actual contribution. He navigated this with the patience of a man who understood that public attention, even when it came from the wrong direction, was better than no public attention, and that the audience he needed to build for serious UFO investigation could not be built without some of them having seen the film.

The media appearances had been building before the film and continued after it. In November 1973 he appeared on The Dick Cavett Show — the same episode that featured Captain Lawrence Coyne describing the Mansfield helicopter incident, which had occurred just weeks before, and two Pascagoula witnesses. The show was a serious engagement with a specific set of cases, conducted on a platform whose guests ranged from

Katharine Hepburn to John Lennon, and Hynek acquitted himself with the careful precision that distinguished his public presentations at their best. Cavett was the right format for him — long-form, intellectually engaged, not interested in spectacle for its own sake. The conversation was about evidence and methodology rather than wonder and sensation, which was Hynek's natural register.

He was a popular and well-compensated speaker on the college circuit throughout the 1970s. He appeared on In Search Of, the Leonard Nimoy-hosted series that explored paranormal and unexplained phenomena and reached an audience of millions weekly. He was a contributing authority to the National Enquirer's Blue Ribbon Panel of scientists, which assessed significant cases and awarded the annual prize for the most scientifically valuable UFO report — the panel that gave the Coyne helicopter case its five-thousand-dollar award in 1973. His involvement with the Enquirer was a source of professional tension that he managed with the pragmatism of someone who understood that reaching the public required going where the public was, even when where the public was happened to be the checkout line at the grocery store.

The swamp gas legacy followed him everywhere, as it always had. It appeared in introductions at conferences and in the first paragraphs of newspaper profiles, in the setups of comedians who found in the phrase a reliable shorthand for government buffoonery. He had long since stopped being visibly bothered by it. What he sometimes said about it — rarely, in the right context, to the right interviewer — was that the swamp gas press conference had been, in retrospect, the most useful thing that had happened to him, because it had made the inadequacy of the official position visible and had eventually forced the kind of public reckoning that made everything he subsequently built possible. This is not revisionism. It is the honest assessment of a man who had spent a decade watching the press conference's consequences ripple outward — the congressional hearings, the Condon Committee, however flawed, the AAAS symposium, the Saturday Evening Post article, CUFOS itself — and had concluded that the moment of public humiliation had been the catalyst for a more honest public conversation than twenty years of quiet professional accommodation had managed to produce.

What the 1970s celebrity did not give him was resolution. He was more visible than he had ever been. The taxonomy was in every encyclopedia. Close Encounters had put his phrase and his name and his image — the pipe, the goatee, the careful astronomer's bearing — into the cultural record in a way that no book could achieve. CUFOS was producing the most rigorous independent UFO investigations ever conducted. And the phenomenon itself remained exactly what it had been on the night he stood in a Michigan swamp in his city shoes trying to find a conventional explanation for something that did not have one: unexplained. The visibility was not the same as an answer. It never had been.

He gave an interview to the Globe and Mail, the Canadian newspaper, at some point in this period — the specific date is not recorded — in which a reporter asked him whether he himself had ever seen a UFO. He said: I have looked up many times, and not seen one thing I would report. But then, I would not report a UFO without a witness. I have seen some strange things — I saw a green fireball over the desert a few years ago — but I would not report it without a witness. This is, in its way, the most complete self-portrait he ever offered. The man who has spent thirty years building the methodology for evaluating other people's reports of extraordinary experiences is holding himself to the same standard he holds everyone else. He wants a witness. He is not less credulous than the profession demands. He is more scrupulous than it has any right to expect.

He is in his late sixties now, a famous astronomer in a field that does not quite recognize the adjective as a compliment. He has built the institution he wanted to build, written the books he wanted to write, made the public case he wanted to make. The question is still open. It will still be open when he dies. He goes back to Evanston and back to the cases, because that is where the answer, if there is one, will be found.

Chapter Thirteen

The Man Himself

On the night of May 5, 1910 — four days after Josef Allen Hynek entered the world — his parents Joseph and Bertha carried their five-day-old son to the roof of their West Side home and held him up to the sky. Halley's Comet was passing close to the Earth that spring, close enough to produce in some quarters genuine terror, and the Hyneks were among the thousands of Chicagoans who had gathered on rooftops and in parks and on the edges of the city to watch. The infant, four days out of the womb, cannot have registered the comet consciously. He was shown it anyway. It was, in some version of the family story, his first astronomical experience.

Seventy-five years and three hundred and fifty-seven days later, on April 27, 1986, J. Allen Hynek died of a malignant brain tumor at Memorial Hospital in Scottsdale, Arizona. Halley's Comet had returned that spring — its first appearance since the night his parents held him up to the Chicago sky. He had wanted to see it. He had been following its approach with the professional attentiveness of a man who had spent his life watching the sky and with the personal investment of someone who understood, or perhaps simply felt, that the comet's return had a significance that pure astronomy could not fully account for. Whether he saw it in the weeks before his death, in the Arizona sky above Scottsdale, is not recorded with certainty. What is recorded is that he was born under Halley's Comet and died under it, bracketed by its appearances in a life that had been, in ways he could not have predicted and could not entirely explain, defined by things in the sky that resisted ordinary explanation.

This chapter is about the life that the previous chapters have not fully told — the private man alongside the public scientist, the family and the faith and the personal history that ran in parallel with the institutional drama, the texture of a daily existence that contained, like most daily existences, more complexity and more ordinary human feeling than any professional biography can fully accommodate.

He married for the first time on December 24, 1932, in Washington County, Arkansas, to Martha Doone Alexander. The marriage lasted approximately seven years and ended in divorce around 1939. He did not speak of it extensively in the available record, and what drove the dissolution is not documented in the sources available to this biography. He was in his late twenties, an instructor at Ohio State, a young astronomer in the early accumulation of a career, and whatever happened between them belonged to a part of his life that he kept private with the same discipline he brought to everything else he wanted to keep private.

In 1942 he married Miriam Curtis — Mimi, always Mimi — fourteen years his junior, in Medina, Ohio. She was, by every account available, the foundation of the life he built afterward. They had five children. Their son Joel became an Oscar-winning visual effects supervisor, the professional heir in a different medium of his father's interest in capturing strange phenomena on camera. Mimi was, in the encyclopedia.com obituary's formulation, a diligent and often unheralded editor and worker behind the scenes — a description that does not begin to account for what it means to be the partner of a man who spent forty years in the professional and social margins of his field, managing the institutional relationships and personal correspondence and logistical complexity of a life organized around a subject that embarrassed most of his colleagues and fascinated the public in ways he sometimes found more burden than blessing. She outlived him by ten years, dying in 1996. The papers are at Northwestern. The full measure of her contribution to his work is there, for anyone who wants to look.

The Rosicrucian question is the one that biographers have circled most carefully, because it complicates the conversion narrative — the story of the skeptic seduced by evidence — that Hynek himself sometimes promoted and that is in many ways the most publicly legible version of his life. Author John Franch, in a Skeptical Inquirer article titled The Secret Life of J. Allen Hynek, argues that the skeptical attitude toward UFOs that Hynek maintained through the Blue Book years was in fact a façade for public consumption. Since his teens, Franch contends, Hynek had been an enthusiastic though closeted student of the occult. The evidence is documented. As a high school student, Hynek spent over one hundred dollars — the equivalent of roughly thirteen hundred dollars today — to purchase Manly Hall's massive illustrated volume The Secret Teachings of All Ages, an encyclopedic survey of Masonic, Hermetic, Qabbalistic, and Rosicrucian philosophy. All my student friends thought I was crazy, he told Vallée years later. Why didn't I buy a motorcycle instead, as they all did? He developed a particular fondness for the writings of the Rosicrucian secret societies and the hermetic philosopher Rudolf Steiner. He later endorsed alleged instances of psychic surgery and psychic photography in ways that his scientific colleagues found indefensible.

What to make of this depends on what you think the conversion narrative was actually explaining. If you believe — as Hynek consistently implied — that he was a pure positivist whose mind was changed by evidence, then the Rosicrucian background is a complication that requires explanation. If you believe — as the evidence increasingly suggests — that he was always looking for the place where conventional knowledge ran out, then there is no contradiction. He told Vallée at some point that he had become an astronomer in order to discover the very limitations of science, the places where it broke down, the phenomena it didn't explain. This is not the stated motivation of a pure empiricist. It is the stated motivation of someone who chose science as the most rigorous available tool for investigating what science itself could not fully account for. The Rosicrucian teenager and the astrophysicist and the UFO investigator are not three different men. They are one man with a consistent underlying question, expressed through the most credible frameworks available to him at each stage of his life.

Ann Eller, who was Hynek's assistant during the final year of his life in Scottsdale, described him in a 2009 Coast to Coast AM interview as a very quiet, private, and soft-spoken man with a twinkle in his eye and a fondness for a good joke or a good story. A true scientist who wanted tangible proof and was troubled by the esoteric aspects of this phenomenon because they complicated the scientific framework he had spent his career building. The esoteric aspects troubled him not because he disbelieved them — his private history with the Rosicrucian material suggests the opposite — but because they were not amenable to the methodology he had developed and trusted, and a man who had spent forty years insisting on methodological rigor could not simply abandon that insistence because it was inconvenient. He held both things simultaneously: the serious scientist who demanded evidence and the private man who had spent his whole life drawn to the territory where evidence ran out. The tension between them was, in the end, productive rather than paralyzing. It kept him honest in both directions.

His relationship with Northwestern deteriorated in the final years before his retirement. The dean of the College of Arts and Sciences said categorically in 1982, speaking for his faculty colleagues: We are not, have not been, and will not be proud of Hynek's UFO affairs. There are many who think that what he's up to has nothing to do with research. This is documented. Hynek had known for years that the faculty regarded his UFO work with varying mixtures of embarrassment and contempt — the Vallée journals record an astronomer friend noting as early as 1972 that The UFO Experience had created antagonism among the Northwestern faculty and made Hynek a controversial figure. He retired in June 1978 at sixty-eight, having run the astronomy department for eighteen years and built it into something considerably stronger than he found it — the television imaging work, the expansion of the department, the legitimate scientific contributions that his Blue Book years had run in parallel with for so long. Northwestern kept the legitimate scientific contributions and held the UFO work at arm's length. This was the institutional arrangement he had been navigating his entire professional life, and he had learned, over those decades, not to expect it to be anything other than what it was.

After retirement CUFOS consumed him. He had hoped the institution would have Northwestern behind it, and it did not. He had hoped it would have serious scientific funding, and the serious scientific funding never materialized at the scale he needed. By 1981 the center had moved from its Chicago offices to his home in Evanston. By 1982 he had disconnected the toll-free hotline that had allowed witnesses to report cases directly. The institution he had built as the anti-Blue Book — properly funded, properly staffed, free of institutional pressure to reach predetermined conclusions — was operating out of a private residence on a shoestring, kept alive by the devotion of the small community of serious investigators who believed in what it was trying to do. He pursued the Corralitos Observatory in New Mexico, persuading Northwestern to donate the now-closed facility to a nonprofit organization he headed — the Corralitos Astronomical Research Association, CARA — which he hoped would provide a physical base for the kind of serious observational work he had always wanted CUFOS to conduct. The observatory was dilapidated and the funding for its restoration never arrived.

In 1984 he moved to Arizona. The move was precipitated by the appearance of what seemed like the funding he had been waiting for — a wealthy Englishman named Jeffery Kaye, a pro-Israeli businessman with homes in three countries, who told Hynek he would back a new International Center for UFO Research in Scottsdale with the resources that CUFOS had never had. Vallée, who visited Scottsdale and recorded his observations in the journals published as Forbidden Science: Volume Three, was skeptical from the first meeting. The people involved were, in his assessment, far from the scientific-minded investigator-researchers Hynek had surrounded himself with throughout his time with CUFOS. One of them,

Vallée winced when present, said that he had jumped on the idea of ufology as a money-making operation when he saw the Meier photographs from Switzerland — the most notorious fraudulent UFO case in the history of the field, which any serious investigator had dismissed years earlier. Hynek, Vallée noted, said nothing. He wanted the funding too badly to say what he knew.

It did not work. The money Kaye had promised was startup funding, not sustained operational funding. The publishing and film projects that were supposed to generate revenue did not materialize. Hynek severed his association with the partners he had come to distrust and was left in Scottsdale with an office, a diminished organization, a relocated family, and the growing awareness that the late-life institutional reinvention he had moved across the country to achieve was not going to happen on the terms he had imagined. Vallée wrote in his journal that Allen himself now felt he had rushed a bit too fast to set up the new center in Arizona under pressure. This is, from Hynek, a considerable admission — he was not a man who rushed, or who admitted to rushing. He had been patient for forty years. The late impatience, the willingness to believe in the funding before the funding was certain, speaks to something in the final years that the public record of the methodical scientist does not fully capture: a man who had been waiting a very long time for the support that would allow the work to be done properly, and who wanted it badly enough, at seventy-four, to trust people he should have evaluated more carefully.

The diagnosis arrived in the midst of this. Prostate cancer first, then a brain tumor discovered as well. The treatment was grueling. He was seventy-five years old, living in a city he had not chosen for its intrinsic qualities but for the promise of resources that had not materialized, separated from the CUFOS community in Evanston that had been the professional family of his post-Blue Book years, dependent on the steadiness of Mimi in a way that was no longer something he could take quietly for granted. He worked when he could. The Hudson Valley UFO Sightings, his final book — co-authored with Philip Imbrogno and Bob Pratt, documenting a wave of sightings in New York's Hudson Valley that began in 1983 — was published in 1987, posthumously. He did not finish it in the form he would have wanted. He finished what he could.

The Chicago Tribune's obituary, published April 30, 1986, quoted a passage from the introduction to his 1977 book. We know so little about the vast universe poised as we are on our tiny vantage point, the Earth. Things far beyond our imagination may be possible. In the meantime, we must satisfy ourselves by studying UFO reports, not UFOs. He added: Reports by people often make mistakes about what we observe. It is, in its compression, a complete statement of where he had arrived — the humility about the scale of what is unknown, the insistence on studying the reports rather than claiming to study the phenomenon itself, the reminder even in a passage about the limits of science that witnesses are fallible and the investigator must account for that. He was honest about the limits of his own conclusions until the end.

His friend and colleague Paola Harris, who had known him since 1980, said after his death that she had cried her eyes out and stopped every part of her research after he died. She stressed that his passing created a void because he was such a highly respected scientist. The void was real — CUFOS continued under Mark Rodeghier's leadership, the case archives were preserved, the methodological standards were maintained, and the work went on. But the specific combination that Hynek represented — the scientific credentials, the Blue Book access, the twenty-one years of institutional memory, the public visibility, the careful honesty — could not be replicated by any combination of successors, because it had been built across an entire lifetime in circumstances that no one would or could deliberately construct.

He was, in a 1973 interview with New Scientist, asked what he would be remembered for. He replied, with the modesty of a man who knew his professional situation precisely: I've never launched any new theories, I've never made any outstanding discoveries. The interviewer, Ian Ridpath, said: I think you'll be remembered not as an astronomer but as the man who made UFOs respectable. Hynek did not argue with this. He had spent twenty-five years making a subject respectable that his profession considered disreputable, and whatever the cost of that in departmental standing and faculty approbation and the wince that crossed certain colleagues' faces when his name came up, it was the work he had chosen. He had chosen it the way you choose things when you are constitutionally incapable of pretending the evidence says something it does not say. He had not chosen it easily or all at once. He had chosen it the way a man who cannot lie eventually, inevitably, chooses the truth — case by case, year by year, at the cost of every comfortable alternative, until the choice was so complete that it had simply become who he was.

He died on a Sunday morning in April, under the comet's return light, in a desert city far from the flat Midwestern skies where the question had found him. The question was still open. It remains open. That was, in the end, the honest answer — and honesty, for J. Allen Hynek, was the only kind of answer he had ever been able to live with.

Chapter Fourteen

The Philosophy of the Unknown

In 1975, J. Allen Hynek published a short essay in the journal Sky and Telescope on the astronomer Johannes Kepler. On its surface the piece was a scholarly appreciation of a seventeenth-century figure — Kepler's mystical vision of the cosmos, his belief that the universe was structured according to mathematical harmonies that were simultaneously scientific and spiritual, his refusal to separate the aesthetic and theological dimensions of the natural world from the mathematical laws he was discovering within it. Beneath the surface it was a self-portrait. Hynek saw in Kepler a kindred spirit — a scientist who understood that the natural world contained more than the instruments of his day could measure, and who was therefore a better scientist for that understanding rather than a worse one. He argued in the essay that there was a tenuous bond linking present-day astronomy's metaphysical thinking and Kepler's brand of metaphysics, with both systems of thought being the repository of fundamental questions not entertained on the present playing field of physical science. He was defending Kepler. He was also, anyone who knew him understood, defending himself.

The essay appeared the same year as The Edge of Reality, and its timing was not accidental. Hynek was sixty-five years old. He had been thinking about the limits of science — its explanatory reach, the territory it could map and the territory it could not — since before he became a scientist, since the teenage boy in Pilsen who spent one hundred dollars he did not have on a book about Rosicrucian philosophy because the astronomy textbooks were not answering the questions that most interested him. He had organized his entire career around the most rigorous available method for investigating the natural world, and that method had led him, case by case and year by year, to the edge of what it could explain. He was now standing at that edge, looking out, and trying to say honestly and precisely what he saw.

What he saw, in the most careful formulation he was able to construct, was this: the UFO phenomenon resisted simple explanation not because the investigators were incompetent or the data was inadequate but because the phenomenon was native to a domain of reality that the instruments and frameworks of materialist science were not designed to reach. This is a precise claim, not a mystical one. It does not assert that the phenomenon is supernatural in the sense of operating outside natural law. It asserts that natural law, as currently understood, is not sufficient — that the phenomenon is evidence of a domain of the natural world that present science has not yet mapped, in the same way that the phenomena of quantum mechanics were evidence, to classical physicists, of a domain that classical mechanics could not account for. He was not abandoning science. He was insisting that science was not yet finished.

The materialist critique of this position, articulated with varying degrees of sophistication by the skeptical community that had been disputing Hynek's work since at least 1966, was that it was unfalsifiable — that a claim about domains beyond the reach of existing instruments cannot be tested and therefore cannot be scientific. Hynek understood this objection and did not dismiss it. He had spent his entire career insisting on falsifiability as the condition of scientific knowledge, and he was not now going to retroactively exempt his own conclusions from that standard. What he argued instead was that the unfalsifiability cut both ways — that the claim that the phenomenon can be explained within existing materialist frameworks was equally untestable, because it required assuming in advance that the instruments adequate to the task existed, and twenty-five years of careful investigation with the best available instruments had not produced an explanation. The methodological discipline he had always practiced — follow the evidence, do not claim more than the evidence supports — pointed him past materialism not because he wanted to go there but because the evidence did not permit him to stay where he was.

His late-life views on the phenomenon went through at least three distinct documented formulations, each more radical than the last, and none of them fully replacing the others — they coexisted in a late-life intellectual position that was provisional and exploratory in ways that his public methodological caution had always tended to conceal. In 1967 he speculated to Vallée that UFOs might be observational devices that are materialized into our world by the denizens of another. This is already a significant departure from the extraterrestrial hypothesis — it does not posit spacecraft traveling from other stars but objects that cross some boundary between domains of reality, materializing in ours for purposes that are not exploratory in the conventional sense. By the early 1980s he was characterizing the phenomenon to the journalist and UFO researcher Jerome Clark as possibly involving elementals — nature spirits in the Rosicrucian and Steinerian sense, beings native to a dimension of the natural world that materialist science does not recognize. In between these formulations he had told interviewers that he thought of UFOs as psychic projections created by an extradimensional intelligence in some parallel reality. These are not the same theory. They share a commitment to the view that the phenomenon is not physical in the way that conventional UFO investigation assumes, and that the framework required to explain it is one that does not yet exist.

The John Franch essay in Skeptical Inquirer argues that these late-life positions reveal that Hynek's UFO investigation was never genuinely empirical — that the Rosicrucian teenager never went away, that the scientist was always in the service of a prior metaphysical commitment, and that his public credibility was therefore based on a misrepresentation of his actual intellectual situation. For Franch, in other words, Hynek wanted to believe in UFOs, and the investigation was shaped by that prior desire rather than by the evidence. The biographer David Halperin offers the more generous and more accurate counterreading: that the question worth asking is not whether Hynek was motivated by a prior attraction to the territory beyond science's reach, but whether his investigation was nonetheless honest. And the evidence that it was honest is overwhelming. He dismissed more cases than he credited. He applied his methodology to his own potential sightings with the same rigor he applied to everyone else's. He resisted the enthusiast community's overclaiming for decades. He held his own late-life positions tentatively, as hypotheses rather than conclusions, and said so explicitly. A man shaped by prior desire in the way Franch describes would not have spent twenty-one years doing the institutional work of a genuine debunker. He would have spent those years finding the phenomenon in every case file rather than in the small percentage of case files that survived rigorous analysis.

The more interesting question, which neither Franch nor Halperin fully pursues, is what the coexistence of the Rosicrucian interest and the scientific rigor actually meant for how Hynek thought. He told Vallée in Colorado, gesturing toward the Rockies and the Great Plains: I have never stopped thinking about what must lie beyond all this. This is the sentence of a man who uses science as a tool for a question that science did not generate — who brought the discipline of astronomy to a question that had its origins not in a seminar room but in a teenage boy's encounter with Manly Hall's survey of hidden knowledge. The discipline is real. The question is real. The fact that they came from different sources does not make either of them less genuine. It makes the combination unusual — more unusual than either a pure empiricist or a pure mystic would have produced — and it may be precisely what allowed him to do what no one else in his position managed to do, which was to follow the evidence honestly into territory that the pure empiricist would have refused to enter and to report what he found with the precision that the pure mystic could not have achieved.

The relationship between consciousness and the UFO phenomenon was the territory where his late thinking was most genuinely exploratory and most genuinely uncomfortable. The abduction reports — which he addressed carefully in The Hynek UFO Report in 1977, noting that CE3 and abduction cases held the most fascination because they brought into focus the concept of other intelligences in space and the possibility of intelligent contact with all that it might imply — raised questions that his methodological framework was not equipped to answer. The details of abduction accounts were almost always obtained through regressive hypnosis, which meant they were products of the subconscious rather than the waking memory, which meant the evidential status of the accounts was fundamentally different from the status of consciously recalled sightings. He could not apply his witness evaluation criteria to hypnotically recovered material with the same confidence he applied them to field reports. He could not cross-check the internal consistency of accounts that emerged from states of consciousness he did not fully understand. He was honest about this. He did not claim that abduction reports were evidence of the same quality as radar-visual cases. He noted them, took them seriously as data of a kind, and acknowledged that they pointed toward questions about the relationship between consciousness and the phenomenon that his existing methodology could not address.

He was increasingly convinced, by the late 1970s, that the phenomenon was interactive in some sense — that it responded to attention, that its presentation varied in ways that seemed calibrated to the observer, that the boundary between the experience of the phenomenon and the phenomenon itself was less clean than the physical evidence approach assumed. This was Vallée's influence, operating at its deepest level. Vallée had argued in Passport to Magonia and in subsequent work that the phenomenon's most consistent characteristic was its evasiveness — its tendency to be exactly as convincing as it needed to be to produce a specific effect in the witness, without being convincing enough to constitute definitive proof.

This was not the behavior of a spacecraft on an exploration mission. It was the behavior of something designed to interact with human consciousness in a particular way, for purposes that might or might not be benign and might or might not be comprehensible within any framework that human thought had yet developed. Hynek found this analysis compelling and could not refute it. He also could not operationalize it — could not turn it into a testable hypothesis that his methodology could address. He held it as a serious unresolved possibility and continued the field investigations that were the only tool he had.

The modern astronomy he described in the Kepler essay — quasars, pulsars, black holes, the exotic physics of the late twentieth century — had become, he argued, a broad playing field for the metaphysicist. He meant this seriously. The universe that physics had revealed by the 1970s was not the mechanistic clockwork that Newton's successors had imagined. It was deeply strange, operating under rules that had no analog in ordinary human experience, producing phenomena that had seemed impossible until the mathematics forced their acceptance. If the universe was strange enough to contain black holes, it was strange enough to contain things that the existing categories of scientific explanation could not accommodate. The question was not whether the phenomenon was real — he was convinced it was — but whether the reality it represented was accessible to the methods of inquiry that had been developed for a different domain of reality. He thought, by the end of his life, that it probably was not. Not because science was inadequate in principle but because the specific science of the late twentieth century had not yet developed the tools for this particular territory, in the same way that Newton's mechanics was not inadequate in principle but was inadequate for the quantum domain. New phenomena required new frameworks. The framework for the UFO phenomenon did not yet exist. His job, as he understood it in the final years, was to document the evidence that such a framework would eventually need to account for, carefully enough that when the framework arrived there would be something for it to explain.

He also thought — and said, in the right contexts, to the right people — that the phenomenon might be telling us something about ourselves that we were not yet equipped to hear. Not a message in the science fiction sense of a decoded transmission. Something more fundamental: that the boundary between observer and observed, between consciousness and the external world, was not as clean as the materialist framework assumed, and that phenomena that appeared to straddle that boundary were evidence of a reality that the hard separation of subject and object could not contain. This is not a mystical claim in the dismissive sense. It is a claim that the philosophy of science of the late twentieth century was actively debating, in the context of quantum mechanics and the measurement problem and the question of what role observation plays in the constitution of physical reality. Hynek was not a philosopher of science. He was an astronomer who had spent forty years watching the evidence point toward a question that philosophy and physics and psychology and anthropology all had pieces of and none of them had whole. He could not answer the question. He could articulate it more precisely than anyone else who had looked at the same evidence, and that precision was its own contribution.

The charge of wishful thinking — Franch's final verdict, that Hynek wanted to believe and the wanting shaped the believing — is the most serious charge that can be leveled at a scientist and the one that most needs to be addressed honestly. The honest answer is that we cannot know, from the outside, the ratio of evidence to desire in any individual's epistemic life. What we can know is the record. And the record shows a man who explained away more sightings than he credited, who held his own conclusions to a higher evidential standard than the enthusiast community demanded, who flagged his uncertainties in print, who told Vallée he would not report a UFO without a witness and meant it, who spent twenty-one years performing the function of a genuine skeptic even as private doubts accumulated, who built an institution designed to test the claims he was drawn to rather than simply to endorse them. That is not the record of wishful thinking. That is the record of a man who was honest about both his desires and the limits of what the evidence could support — which is, in the end, as close to scientific virtue as any human investigator is likely to get.

He once told an interviewer, in the last decade of his life: I have never stopped thinking about what must lie beyond all this. He was gesturing at the landscape — the mountains, the plains, the visible world that his instruments had spent a career measuring. But he might as well have been gesturing at the entire territory his life had crossed: the observatory at Yerkes, the Blue Book files in Dayton, the marshy ground outside Dexter, Michigan, the AAAS symposium in the Boston snow, the Mobile airplane hangar where Spielberg's film was becoming real, the CUFOS archive in Evanston, the Scottsdale hospital room at the end. All of it, the whole forty-year voyage, had been an attempt to think rigorously about what lies beyond the boundary of the known. He never reached that boundary. He never expected to. He kept moving toward it until he could not anymore, holding the question open with the same discipline he had brought to every other scientific problem in his life, refusing the comfort of a premature answer with the same integrity that had made him, at considerable cost, unable to keep saying what he did not believe.

That is the philosophy of J. Allen Hynek, stated in its simplest form: that the unknown deserves the same rigorous, honest, patient attention that the known has received. That ridicule is not part of the scientific method. That the question is real. And those real questions, however long they remain open, are worth the cost of asking them honestly.

Epilogue-His Shadow

On December 16, 2017, the New York Times published a story under the headline Glowing Auras and Black Money: The Pentagon's Mysterious U.F.O. Program. The reporters — Helene Cooper, Leslie Kean, and Ralph Blumenthal — had obtained documents confirming the existence of the Advanced Aerospace Threat Identification Program, a secret Pentagon initiative funded with $22 million secured by Senator Harry Reid of Nevada, which had operated from 2007 to 2012 investigating military encounters with objects that Navy pilots were reporting in controlled airspace with increasing frequency. The story ran with declassified video footage — three clips from Navy aircraft, showing objects moving in ways that the pilots could not explain and that no known aircraft could perform — which the Pentagon later officially acknowledged as genuine. Luis Elizondo, the career intelligence officer who had run AATIP, had resigned his position in protest, writing in his resignation letter to Secretary of Defense James Mattis that the government was not taking the issue seriously. He went public the same day the Times story ran.

J. Allen Hynek had been dead for thirty-one years.

Everything that has followed the 2017 Times story — the congressional hearings, the Pentagon office, the whistleblower testimony, the NASA study panel, the shift in the national conversation from fringe to front page — is built on a foundation that Hynek spent forty years constructing without knowing whether anyone would ever build anything on it. His taxonomy is in current use. The term close encounter, which he coined as a methodological category in The UFO Experience, is in every congressional report and Pentagon briefing document. The vocabulary of serious UAP investigation — residuals, physical trace evidence, electromagnetic effects, radar-visual cases, witness credibility assessment — is his vocabulary, developed across twenty-one years of Blue Book consultancy and fifteen years of independent investigation, now being used by military pilots and congressional staffers and Pentagon officials who may or may not know where it came from.

The institutional history that the 2017 moment revealed was, to anyone who had read Hynek carefully, entirely familiar. AATIP had operated in secret, without congressional oversight, for years — a hidden program investigating cases that military pilots were reporting and that official channels were systematically not equipped to handle. The Bolender memo of 1969, which Hynek did not know about when Blue Book closed, had said exactly this: that the cases most warranting investigation were not going through Blue Book at all, but through other channels, through procedures designed for this purpose, the nature of which was not explained. What the 2017 revelations suggested was that those other channels had never stopped running. The gap between the public investigation — Blue Book, then nothing, then AATIP briefly, then AARO — and whatever was actually happening in the classified programs was not new. It was fifty years old. Hynek had suspected its existence for most of his professional life without being able to prove it.

The congressional engagement that followed the 2017 story moved faster than anything Hynek had managed to produce in forty years of effort. In May 2022, the House Intelligence Subcommittee held the first public congressional hearing on UFOs in over fifty years — the previous one having been the Roush Symposium of 1968, at which Hynek had testified as a private scientist and called for exactly the kind of independent investigation that the 2022 hearing was now, belatedly, beginning to demand. In July 2022, the Pentagon established the All-domain Anomaly Resolution Office — AARO — with a broader mandate than any predecessor, covering air, sea, space, and transmedium environments. In July 2023, the House Oversight Subcommittee held what many observers characterized as the most consequential congressional hearing on the subject in history, featuring sworn testimony from David Grusch, a former Air Force intelligence officer and representative to the UAP Task Force, who stated under oath that the United States government had recovered non-human craft and biological materials and had been operating programs to study them without congressional knowledge or authorization for decades.

Grusch's testimony was specific, credentialed, and made under oath before a congressional committee. It was also — and this is the point that Hynek's ghost, if such a thing existed, would have appreciated with exquisite irony — made without physical evidence that could be publicly examined. Grusch was reporting what he had been told by individuals with long-standing track records of legitimacy and service, whose own testimony he had spent four years attempting to corroborate. He had not personally handled the recovered materials. He was a credible witness reporting credible testimony from other credible witnesses, which is precisely the category of evidence that Hynek spent his career arguing deserved rigorous investigation rather than institutional dismissal. The skeptics dismissed it — AARO's director Sean Kirkpatrick testified that his office had found no credible evidence of extraterrestrial activity — and Grusch replied that Kirkpatrick's statement was not accurate. The pattern was perfect: an official body finding nothing, an insider saying the official body is not looking in the right place, the public record split between the two positions with no resolution in sight.

Grusch's own language is a precise echo of the methodological care that Hynek practiced. Asked whether he was saying the objects were extraterrestrial, he said: I like to use the term non-human. I don't like to denote origin. It keeps the aperture open scientifically. He said it in a congressional hearing, under oath, in 2023. Hynek had been saying the same thing — in different words, with different evidence, against different institutional resistance — since at least 1966. Do not claim more than the evidence supports. Do not close the aperture prematurely. Follow the evidence to where it actually goes rather than to where the institutional framework needs it to go. This is Hynek's methodology. It is also, apparently, what a serious intelligence officer brings to the same material fifty years later when the institutional pressure is different but the epistemological problem is identical.

Kirkpatrick's characterization of UAP advocacy as basically a religion, a religious belief that transcends critical thinking and rational thought is an almost verbatim repetition of the position that Hynek spent his entire career arguing against. Ridicule is not part of the scientific method. The dismissal of witness testimony from credentialed observers on the grounds that the phenomenon they are reporting cannot exist is not analysis. It is the decision, made before the investigation begins, that the investigation will find nothing. Kirkpatrick did not invent this position. He inherited it from a long institutional tradition that runs from the Robertson Panel through the Condon Report through Blue Book's closure — the tradition of official bodies whose function was management of public perception rather than scientific investigation of an anomalous phenomenon. Hynek recognized this tradition in 1953. He spent the rest of his life trying to do something about it. In 2023 it was still running, still performing the same function, still producing the same conclusions from the same predetermined posture.

CUFOS continues. Mark Rodeghier, who succeeded Hynek in 1986, has run the organization for nearly four decades, maintaining the case archives, preserving the methodological standards, publishing the International UFO Reporter, and ensuring that the database Hynek and Vallée built in the 1960s is not lost. The organization operates on a shoestring — the serious scientific funding that Hynek wanted and never fully obtained has still not materialized — but it exists, and the archives exist, and the 701 Blue Book unknowns and the thousands of subsequent carefully investigated cases are preserved and available to researchers who want to examine them. NASA's 2023 UAP study panel, which was the first time a major government science agency had formally engaged with the subject since the Condon Report, cited CUFOS materials in its background research. The foundation Hynek built is being used by people who are finally beginning to understand what it was built for.

The shift from UFO to UAP — Unidentified Aerial Phenomena, then Unidentified Anomalous Phenomena as the definition expanded to include transmedium objects — is partly semantic and partly substantive. Semantically it was designed to separate the subject from fifty years of cultural baggage, the flying saucers and little green men and supermarket tabloids that Hynek spent his career distinguishing serious investigation from. Substantively it reflects exactly the methodological point Hynek always insisted on: the phenomenon should be described by what it is observed to do, not by what the observer assumes it to be. An unidentified anomalous phenomenon is precisely what Hynek called a residual — an observation that survives rigorous application of conventional explanations and remains genuinely unexplained. The new terminology is his old methodology, finally finding its way into official language.

His name has not become famous in the way that the current moment might suggest it should be. Most of the congressional representatives who have led the UAP hearings cannot identify the origin of the classification system they are routinely invoking. Most of the journalists covering the UAP beat have read the Times stories and the Grusch testimony and the AARO reports without reading The UFO Experience. Most of the military pilots filing UAP reports are using reporting protocols whose intellectual genealogy runs back to the CUFOS methodology of the 1970s without being aware of that genealogy. He is the foundation that was poured before the building went up, which is to say he is present in everything and visible in almost nothing. This would not have surprised him. He was not building toward celebrity. He was building toward the moment when the question would be taken seriously, and that moment has arrived, and he is not here to see it.

What would he have made of it? The question is worth holding for a moment before answering. He would have been gratified by the congressional engagement, which is what he called for in April 1966 and what did not happen seriously until May 2022. He would have been gratified by the pilots — credentialed military observers filing official reports about unexplained phenomena, which is exactly the category of evidence he argued for decades deserved rigorous investigation rather than institutional dismissal. He would have been gratified by the shift in scientific respectability, by NASA's study panel, by the academic papers appearing in peer-reviewed journals where such papers were previously unpublishable. He would have been cautious about the whistleblower claims, not because he disbelieved them but because his methodology demanded evidence rather than testimony, and the physical evidence that Grusch's testimony pointed to has not been publicly produced. He would have wanted to examine it. He would have known, from fifty years of watching institutions manage this subject, that the gap between what the insiders say and what the public evidence shows is precisely the gap that needs to close before any conclusion is scientifically warranted.

He would also, probably, have been unsurprised by Kirkpatrick's dismissal and by AARO's hedged acknowledgments and by the pattern of institutional resistance that has accompanied every step of the current moment's progress. He had watched that pattern operate for forty years. It is not a conspiracy. It is the normal behavior of institutions confronted with material that their existing frameworks cannot accommodate and whose acknowledgment carries costs they are not yet ready to pay. The Robertson Panel did it in 1953. The Condon Committee did it in 1968. AARO is doing it now. The pattern does not change because the institution changes. It changes when the evidence becomes sufficient that the cost of continuing to suppress it exceeds the cost of acknowledging it, and Hynek understood that reaching that threshold was not primarily a scientific problem. It was a political and cultural problem, and scientific humility — the willingness to say I don't know rather than nothing to see here — was the prerequisite for any genuine progress.

He said, in a 1975 paper delivered to the Joint Symposium of the American Institute of Aeronautics and Astronautics, that phenomena which require explanation beyond present theoretical capabilities include the reported ability of UFOs to materialize and dematerialize, the associated electromagnetic and physiological effects, the reported psychic effects on percipients, and the structured appearance and behavior of reported occupants. He was describing, in precise technical language, a body of evidence that he had accumulated across twenty-five years of investigation and that he could not explain within any existing scientific framework. He did not know what explained it. He knew what the evidence showed and he said so. Ridicule is not part of the scientific method. The question is real. The investigation has been inadequate. These statements were true in 1953 and true in 1975 and true in 1986 when he died and true today.

The most honest thing that can be said about the current UAP moment, from the vantage point of Hynek's life, is that it is the beginning of the investigation he spent forty years trying to start. Not the conclusion. Not the revelation. The beginning. The institutional acknowledgment that something is genuinely happening that official science has not adequately addressed. The first serious attempt, under sustained congressional pressure, to build the kind of reporting infrastructure and analytical apparatus that Hynek built in miniature at CUFOS and that the Air Force never built in twenty-one years of Blue Book. The first moment in fifty years when a military pilot can file a UAP report without career consequences — a change so basic, so obviously necessary, so long in coming that its arrival feels both overdue and somehow fragile.

The question remains open. It has always remained open. That is the honest answer, and Hynek spent his life learning that the honest answer, however unsatisfying, was the only answer worth giving. He was born under Halley's Comet and died under it, bookended by the same astronomical visitor in a life that had been defined by things in the sky that returned without explanation. He taught us how to look. He taught us what to say when what we saw didn't fit the categories we had. He taught us that the question itself was worth the cost of asking honestly, and that the cost was real, and that paying it was what a scientist who took his obligations seriously was obligated to do.

His shadow falls across every congressional hearing, every Pentagon report, every Navy pilot who has filed an honest account of something in the sky that did not fit the available categories. He is the foundation that was poured before the building went up. He did not live to see the building. He built the foundation anyway — carefully, precisely, honestly, at considerable cost — because that was what the evidence required of him, and he was, in the end, a man who could not do less than what the evidence required.

Chapter Fifteen

Bibliography

Prologue

Hynek, J. Allen. *The UFO Experience: A Scientific Inquiry.* Chicago: Henry Regnery, 1972.

Vallée, Jacques. *Forbidden Science: Journals 1957–1969.* Berkeley: North Atlantic Books, 1992.

Clark, Jerome. *The UFO Encyclopedia.* 3rd ed. Detroit: Omnigraphics, 2018.

Fuller, John G. *Incident at Exeter.* New York: G. P. Putnam's Sons, 1966.

United States Congress. House. *Unidentified Flying Objects: Hearing before the Committee on Armed Services.* 89th Cong., 2nd sess., April 5, 1966.

Chapter One

Hynek, J. Allen. *The UFO Experience: A Scientific Inquiry.* Chicago: Henry Regnery, 1972. *(Autobiographical passages regarding early career and scientific formation.)*

Schmitt, Donald R. and Carey, Thomas J. *Inside the Real Area 51.* Franklin Lakes: New Page Books, 2013. *(Background on postwar Air Force scientific recruitment.)*

Vallée, Jacques. *Forbidden Science: Journals 1957–1969.* Berkeley: North Atlantic Books, 1992. *(Vallée's account of Hynek's character and background as related in conversation.)*

Clark, Jerome. *The UFO Encyclopedia.* 3rd ed. Detroit: Omnigraphics, 2018. *(Biographical summary of Hynek's early career.)*

Dobrovolny, Jerry. "J. Allen Hynek: The Man Behind the Myth." *International UFO Reporter* 11, no. 5 (1986). *(Obituary and career summary published shortly after Hynek's death.)*

University of Chicago. *Dissertations in Astrophysics, 1930–1940.* Chicago: University of Chicago Library Archives. *(For verification of dissertation details.)*

Johns Hopkins Applied Physics Laboratory. *History of the Proximity Fuze Program, 1940–1945.* Laurel: APL Archives. *(For verification of Hynek's wartime role.)*

Chapter Two

Hynek, J. Allen. *The UFO Experience: A Scientific Inquiry.* Chicago: Henry Regnery, 1972. *(Autobiographical reflections on scientific training and methodology.)*

Hynek, J. Allen. *The Astrophysics of Variable Stars.* Unpublished lecture notes, Northwestern University Archives, Evanston, Illinois. [*Verify availability.*]

Kevles, Daniel J. *The Physicists: The History of a Scientific Community in Modern America.* New York: Alfred A. Knopf, 1978. *(Context for mid-century American academic science and its culture.)*

DeVorkin, David H. *Science with a Vengeance: How the Military Created the US Space Sciences After World War II.* New York: Springer, 1992. *(Context for wartime and postwar scientific culture and the relationship between academic and applied science.)*

Vallée, Jacques. *Forbidden Science: Journals 1957–1969.* Berkeley: North Atlantic Books, 1992. *(Vallée's portrait of Hynek's intellectual character and methodology as observed firsthand.)*

Clark, Jerome. *The UFO Encyclopedia.* 3rd ed. Detroit: Omnigraphics, 2018. *(Career summary and bibliographic sources.)*

Northwestern University Archives. *Faculty Records: J. Allen Hynek, 1960–1986.* Evanston: Northwestern University Library. [*Verify accessibility and holdings.*]

Ohio State University Archives. *Department of Astronomy Records, 1936–1960.* Columbus: Ohio State University Library. [*Verify accessibility and holdings.*]

Chapter Three

Ruppelt, Edward J. *The Report on Unidentified Flying Objects.* Garden City: Doubleday, 1956. *(Ruppelt's firsthand account of the early Air Force UFO projects, including Sign and Grudge, remains the most authoritative insider narrative of the institutional culture.)*

Hynek, J. Allen. *The UFO Experience: A Scientific Inquiry.* Chicago: Henry Regnery, 1972. *(Hynek's retrospective account of his early consulting role and evolving attitudes.)*

Vallée, Jacques. *Forbidden Science: Journals 1957–1969.* Berkeley: North Atlantic Books, 1992. *(Vallée's account of Hynek's description of the early project years as related in conversation.)*

Clark, Jerome. *The UFO Encyclopedia.* 3rd ed. Detroit: Omnigraphics, 2018. *(Summary of Projects Sign and Grudge with source bibliography.)*

Dolan, Richard M. *UFOs and the National Security State: Chronology of a Cover-Up, 1941–1973.* Charlottesville: Hampton Roads, 2002. *(Detailed account of early Air Force UFO projects and institutional politics, including the Estimate of the Situation.)*

United States Air Force. *Projects Sign, Grudge, and Blue Book: Declassified Records.* National Archives and Records Administration, Record Group 341. Washington: NARA. *(Primary source material for Air Force project documentation.)*

Jacobs, David Michael. *The UFO Controversy in America.* Bloomington: Indiana University Press, 1975. *(Academic history of American UFO investigation and its institutional context.)*

Gross, Loren E. *UFOs: A History.* Fremont: privately published, 1982–2005. *(Detailed chronological documentation of cases and institutional history across multiple volumes.)*

Chapter Four

Ruppelt, Edward J. The Report on Unidentified Flying Objects. Garden City: Doubleday, 1956. Ruppelt's firsthand account of Blue Book operations and institutional culture under his directorship, including his notes on Hynek's role and the dynamics of the Robertson Panel discussions.

Durant, Frederick C. Report of Meetings of Scientific Advisory Panel on Unidentified Flying Objects. Washington: Central Intelligence Agency, 1953. The Robertson Panel report, declassified and available through CIA FOIA releases. Confirms panel composition, four-day timeline, findings, and recommendations regarding public debunking.

Hynek, J. Allen. The UFO Experience: A Scientific Inquiry. Chicago: Henry Regnery, 1972. Hynek's retrospective account of the Blue Book years, his evolving private doubts, and his characterization of the Robertson Panel's methodology as inadequate.

Vallée, Jacques. Forbidden Science: Journals 1957–1969. Berkeley: North Atlantic Books, 1992. Vallée's journals contain detailed accounts of Hynek's private assessments of Blue Book and his institutional situation, as related in conversation from 1963 onward.

Jacobs, David Michael. The UFO Controversy in America. Bloomington: Indiana University Press, 1975. Academic analysis of the Robertson Panel, its recommendations, and their consequences for official investigation and public discourse.

Dolan, Richard M. UFOs and the National Security State: Chronology of a Cover-Up, 1941–1973. Charlottesville: Hampton Roads, 2002. Detailed account of the institutional politics of Blue Book in the post-Ruppelt years, including the Quintanilla period.

Clark, Jerome. The UFO Encyclopedia. 3rd ed. Detroit: Omnigraphics, 2018. Entries on Blue Book, Robertson Panel, Quintanilla, and Hynek's documented role and assessments across the Blue Book period.

United States Air Force. Project Blue Book Declassified Files. National Archives and Records Administration, Record Group 341. Washington: NARA. Primary source case files and administrative records covering 1952–1969.

Hynek, J. Allen. Private correspondence and notes, 1952–1969. CUFOS Archives, Chicago. Primary source for Hynek's private assessments diverging from public Blue Book conclusions. Verify specific holdings and accessibility with CUFOS directly.

Smithsonian Astrophysical Observatory. Operation Moonwatch Records, 1956–1960. Smithsonian Institution Archives, Washington. Documents Hynek's role directing the satellite tracking program during the International Geophysical Year and the Sputnik period.

Life Magazine. Cover and feature coverage of Operation Moonwatch and Sputnik. October 1957. Documents Hynek's public visibility as a conventional astronomer during the same period he was conducting Blue Book consultancy work.

Chapter Five

Hynek, J. Allen. The UFO Experience: A Scientific Inquiry. Chicago: Henry Regnery, 1972. Contains Hynek's retrospective account of the Socorro case, his assessment of Zamora's credibility, and his analysis of the Portage County case including a partial transcript of Quintanilla's witness interview, which Hynek uses to document the institutional bad faith of the Blue Book investigation.

Vallée, Jacques. Forbidden Science: Journals 1957–1969. Berkeley: North Atlantic Books, 1992. Vallée's contemporaneous journal entries from his Northwestern years provide the most detailed available record of his collaboration with Hynek and Hynek's private thinking during this period. Essential primary source for the invisible college formation and the Hynek-Vallée intellectual relationship.

Vallée, Jacques. The Invisible College: What a Group of Scientists Has Discovered About UFO Influences on the Human Race. New York: Dutton, 1975. The book that named and described the network of credentialed scientists privately engaged with the phenomenon. Draws on the same milieu documented in the Forbidden Science journals.

Enigma Labs. Socorro UAP Incident (Lonnie Zamora Incident). enigmalabs.io. Detailed case summary drawing on primary documents including Hynek's September 1964 letter to Menzel calling Socorro a potential Rosetta Stone case and describing Zamora as the most unimpeachable witness he had encountered. Confirms Hynek's follow-up investigation date of August 15, 1964.

Black Vault Case Files. Project Blue Book: Socorro, New Mexico UFO Landing, April 24, 1964. theblackvault.com. Reproduces CIA Studies in Intelligence brief by Quintanilla confirming his assessment of the case as one of the best documented on record. Confirms Hynek's belief in Zamora's account and his statement that the case represents a most real phenomenon of undetermined origin.

Randle, Kevin D. Encounter in the Desert: The Case for Alien Contact at Socorro. New Page Books, 2016. The most thorough book-length treatment of the Socorro case, drawing on primary documents and interviews. Useful for confirming witness credibility assessments and the scope of the investigation.

Clark, Jerome. The UFO Encyclopedia. 3rd ed. Detroit: Omnigraphics, 2018. Entries on Socorro, the invisible college, Vallée's career, and the Blue Book period from 1963 onward.

United States Air Force. Project Blue Book Declassified Files, Case No. 8766 (Socorro, NM, April 24, 1964). National Archives and Records Administration, Record Group 341. Washington: NARA. Primary source documentation of the Air Force's own investigation, listed as Unknown.

Hynek, J. Allen. Private correspondence with Donald Menzel, September 1964. CUFOS Archives, Chicago. Contains the Rosetta Stone characterization and the most unimpeachable witness assessment. Verify specific holdings and accessibility with CUFOS directly.

Biography.com. J. Allen Hynek: Project Blue Book, Quotes and Facts. Confirms the invisible college formation at Northwestern in the 1960s and its specific connection to the Socorro case as a catalyst for Hynek's more open private discussions with colleagues.

Chapter Six

Wikipedia. Michigan 'Swamp Gas' UFO Reports. en.wikipedia.org. The most comprehensive single summary of the chronology, incorporating primary source documents and contemporaneous press accounts. Confirms press conference location as Detroit Press Club, March 25, 1966, sixty press members attending. Confirms congressional hearing date and Hynek's statement. Confirms the 1984 Omni magazine interview in which Hynek called the incident a turning point.

Bentley Historical Library, University of Michigan. Flying Saucers and Swamp Gas. bentley.umich.edu. Institutional archive account drawing on primary documents from the University of Michigan, the Ford Presidential Library, and Michigan congressional representatives' files. Confirms chronology, Harvey's role, and the political sequence from swamp gas to the Condon Committee.

ClickOnDetroit. 'I'll Believe This to the Day I Die': The Michigan UFO Craze of 1966. clickondetroit.com. Includes Sheriff Douglas Harvey's 2006 extended recollection of driving Hynek to the Mannor farm, Hynek's admission of genuine uncertainty in the car, the private phone call to Washington, and his verbatim account of Hynek's transformation from 'I don't know' to 'it's swamp gas.' Essential primary testimony.

ParaRational. The 1966 Dexter Michigan UFO Sighting at 60: How 'Swamp Gas' Changed UFO History Forever. pararational.com. Confirms Hillsdale details including 87 witnesses, football-shaped object per Blue Book's own file, Van Horn's firsthand observation, and the timeline of the sightings through 5:10 a.m.

Michigan Today, University of Michigan. Ann Arbor vs. the Flying Saucers. michigantoday.umich.edu. Includes Harvey's direct quote about the phone call, confirms Hynek's awareness that he was in trouble when he saw reporters underlining 'swamp gas,' and provides contemporaneous press reaction context.

The Black Vault. The Road to a Congressional UFO Hearing, 1960s Style. theblackvault.com. Documents the Ford letter of March 28, 1966, the congressional hearing of April 5, 1966, Hynek's statement calling for a civilian scientific panel, and Ford's subsequent press release of April 21, 1966, accepting the Condon Committee as a step in the right direction. Draws on documents from the Gerald Ford Presidential Library.

United States Congress. House Committee on Armed Services. Unidentified Flying Objects: Hearing. 89th Congress, 2nd Session, April 5, 1966. Available via Internet Archive (ia600300.us.archive.org). Primary source transcript of the congressional hearing confirming Hynek's presence alongside Secretary Brown and Quintanilla, the reading of his statement, and the committee's findings.

Hynek, J. Allen. Are Flying Saucers Real? Saturday Evening Post, December 17, 1966. Hynek's first major public break with institutional orthodoxy, published under his own name in a mass circulation magazine.

Available through the Ann Druffel Special Collection of the Expanding Frontiers Archive.

Hynek, J. Allen. The UFO Experience: A Scientific Inquiry. Chicago: Henry Regnery, 1972. Contains retrospective account of the Michigan press conference and Hynek's subsequent shift, including the Omni characterization of Michigan as a turning point.

Air and Space Forces Magazine. USAF and the UFOs. airandspaceforces.com. Confirms the Hillsdale detail of 87 student witnesses and Ford's demand for a full investigation, as well as the one hour and twenty minute duration of the congressional hearing.

Gerald Ford Presidential Library. Ford letter to House Armed Services and Science Committees, March 28, 1966. fordlibrarymuseum.gov. Primary source. Ford explicitly describes Hynek as having dismissed the sightings as the product of college student pranks, swamp gas, or an impression created by the rising crescent moon and Venus — confirming the breadth of the press conference's dismissals beyond swamp gas alone.

Date discrepancy note: The congressional hearing is listed as April 5 in the Black Vault documentation and the hearing transcript itself, which states Tuesday, April 5, 1966, at 10:35 a.m. (April 5, 1966 was indeed a Tuesday). Wikipedia cites April 6. April 5 should be used as the confirmed date. The discrepancy may originate from a separate April 6 Air Force announcement about the forthcoming Condon Committee study.

Chapter Eight

United States Congress. House Committee on Science and Astronautics. Symposium on Unidentified Flying Objects: Hearings. 90th Congress, 2nd Session, July 29, 1968. Washington: Government Printing Office, 1968. Primary source transcript of the Roush Symposium, available through NICAP archives (nicap.org). Confirms Hynek's opening statement that he is speaking as a private citizen rather than an Air Force representative, the order of presenters, Rumsfeld's introduction of Hynek, and Hynek's remarks about a large number of privately interested scientists.

Hynek, J. Allen. Twenty-One Years of UFO Reports. Paper presented at the 134th Meeting of the American Association for the Advancement of Science, UFO Symposium, Boston, December 27, 1969. Published in Sagan and Page, UFOs — A Scientific Debate (1972). Available through CUFOS archives (cufos.org/hynek-biography/articles-and-talks/). Confirms symposium dates, the heavy snowstorm, Hynek's title, and the direct quotations regarding the improved intellectual climate and his reasons for delayed public advocacy.

Sagan, Carl, and Thornton Page, eds. UFOs — A Scientific Debate. Ithaca: Cornell University Press, 1972. The published proceedings of the AAAS symposium. Reviewed in Science 180 (1973). Five of six symposium scientists supported serious investigation; only Sagan was more cautious.

Hynek, J. Allen. The UFO Experience: A Scientific Inquiry. Chicago: Henry Regnery, 1972. Reprinted by Da Capo Press, 1998. ISBN 978-1-56924-782-2. The primary text for this chapter. 276 pages plus plates. Introduces the six-fold classification system in full. Reviewed in Science 177 (1972) by Bruce C. Murray under the title Reopening the Question.

Enigma Labs. Hynek's Close Encounters Scale. enigmalabs.io. Confirms the taxonomy first outlined at the AAAS symposium December 1969 and fully developed in The UFO Experience (1972). Confirms Spielberg payment details: $1,000 for title use, $1,000 for book stories, $1,500 for three days consulting. Confirms the film was originally titled Watch the Skies. Source cited as Hynek's biographer Mark Connell.

Wikipedia. Close Encounter. en.wikipedia.org. Confirms the six-fold classification, the 500-foot proximity threshold for close encounters, Hynek's deliberate choice of animate beings rather than extraterrestrials, and the cameo in the 1977 Spielberg film. Notes that CE4 and CE5 were added by others after Hynek's death and were not part of his original system.

CUFOS. Classification Systems. cufos.org/types-of-ufos/classification-systems/. Authoritative source for Hynek's original categories as adopted and used by CUFOS from its inception. Confirms the definitions of all six categories.

McCarthy, Paul. UFOs and the Behavioral Scientist: An Analysis of UFO Studies Since World War II. Unpublished dissertation. Chapter VI: The 1969 AAAS UFO Symposium. Available at files.ncas.org/mccarthy/chap06.html. The most detailed historical account of the AAAS symposium's organization, opposition from Condon and Menzel, and the limited scientific repercussions that followed. Confirms McDonald missed the first day due to the Boston snow. Confirms the assessment that the UFO situation remained much the same before and after the symposium.

Wikipedia. Condon Committee. en.wikipedia.org. Confirms the universities that declined to host the Colorado study including Harvard, MIT,

Berkeley, and North Carolina. Confirms Hynek and McDonald both suggested their own campuses and were rejected as too closely allied with one position. Confirms Hynek's characterization of Condon's introduction as singularly slanted and his finding that more than a quarter of cases remained unexplained.

Vallée, Jacques. Forbidden Science: Journals 1957–1969. Berkeley: North Atlantic Books, 1992. Covers the period through the AAAS symposium and Hynek's public transition from consultant to independent investigator.

Note on the AAAS symposium date: The symposium took place December 26 and 27, 1969. Some sources cite December 27 only, which is the date of Hynek's specific presentation. The full symposium ran both days. The CUFOS archive confirms the 134th Meeting of the AAAS.

Chapter Nine

Hynek, J. Allen. Unusual Aerial Phenomena. Journal of the Optical Society of America 43, no. 4 (April 1953): 311–314. The primary source for this chapter's opening passage. Verbatim quote confirmed across multiple independent sources. The article title Unusual Aerial Phenomena is significant in retrospect as a precursor to the UAP terminology now in official use.

Hynek, J. Allen. The UFO Experience: A Scientific Inquiry. Chicago: Henry Regnery, 1972. The primary text for the methodology sections. Contains Hynek's full exposition of witness evaluation criteria, the taxonomy rationale, physical trace evidence categories, EM effects documentation, and the residuals concept.

Hynek, J. Allen. The Hynek UFO Report. New York: Dell, 1977. Contains retrospective assessment of Blue Book methodology, the statement that he enjoyed his debunker role and that debunking was what the Air Force expected. Essential for the insider view of institutional pressures on the investigation.

Hynek, J. Allen. The UFO Gap. Playboy, December 1967. Contains the verbatim statement about what a proper investigation would produce within a year of initiation: photographs, movies, spectrograms, plaster casts of indentations, and detailed measurements. Available through the Expanding Frontiers Archive.

History Channel. Meet J. Allen Hynek, the Astronomer Who First Classified UFO Close Encounters. history.com. Confirms Joel Hynek's statement about his father's emphasis on keeping an open mind, and Joel's role recording witness interviews as a young ham-radio operator. Attributes the Spielberg payment figures to reporter Mark O'Connell.

Collins, Curt. Dr. Hynek and the UFO Photo Investigation of 1967. Blue Blurry Lines weblog. Comprehensive documentation of the Grant Jaroslaw Michigan photo case, Hynek's denial of access to the original prints, his hedged public assessment, the resulting misleading headlines, and the nine-years-later confession letter. Characterizes Hynek's role as an unwitting secondary dupe. Essential for the photographic methodology section.

Wikipedia. J. Allen Hynek. en.wikipedia.org. Confirms the statement to a Globe and Mail reporter that Hynek had never seen a UFO he would report without a witness, and the context in which it was made alongside the Vallée stereoscopic photograph discrepancy. Confirms the 1953 JOSA article exact title, date, and verbatim quote.

Thinkaboutit.site. The Hynek Classification System. Documents the full physical trace evidence and medical record evidence categories as defined in Hynek's system, including compressed vegetation, broken branches, ground imprints, and medical verification of physiological effects as categories of physical evidence.

Vallée, Jacques. Forbidden Science: Journals 1957–1969. Berkeley: North Atlantic Books, 1992. Documents the computerized analysis of Blue Book cases that confirmed the non-random distribution of residuals. Also contains the March 1968 journal entry noting that Hynek's colleagues' attitude toward him was changing to the point of contempt, confirming the professional costs were real.

CUFOS. Hynek Archives. cufos.org. Documents the 1952 informal survey of astronomers at the American Astronomical Society meeting in Vancouver, which was Hynek's first systematic attempt to assess scientific opinion on UFOs. Confirms the existence of correspondence, topic files, case notes, and witness interviews preserved from the Hynek collection. Confirms the 1968 exchange with Colonel Sleeper as Hynek's most in-depth in-house critique of Blue Book methodology.

UAP Wiki. J. Allen Hynek. uap-wiki.org. Confirms that by the early 1960s, per Jerome Clark, Hynek's apparent turnaround on the UFO question was an open secret. Also confirms Hynek's own retrospective assessment that his personality — careful, diplomatic, non-confrontational —

was a factor in the Air Force keeping him as a consultant for over two decades despite his growing divergence from institutional conclusions.

Hynek, J. Allen. Axioms for a Developing Science of UFOs. Paper presented at the Joint Symposium of the American Institute of Aeronautics and Astronautics, Los Angeles, 1975. Contains the quoted passage on phenomena requiring explanation beyond present theoretical capabilities, including EM effects, psychic effects on percipients, and the materialization and dematerialization problem. Documents his late-career shift toward the extradimensional hypothesis.

Chapter Ten

Vallée, Jacques. Forbidden Science: Journals 1957–1969. Berkeley: North Atlantic Books, 1992. Volume 1 of the published journals. The most detailed available account of the Hynek-Vallée intellectual relationship as it developed in real time, including the erased Paris Observatory tracking tapes, the early Northwestern collaboration, and Vallée's growing divergence from the extraterrestrial hypothesis.

Vallée, Jacques. Forbidden Science: Journals 1970–1979. Volume 2. Berkeley: North Atlantic Books, 2009. Covers the period of The Edge of Reality, the CUFOS years, and the developing interdimensional hypothesis. Documents the continuing Hynek-Vallée conversations through the 1970s.

Hynek, J. Allen, and Jacques Vallée. The Edge of Reality: A Progress Report on Unidentified Flying Objects. Chicago: Henry Regnery, 1975. Reprinted by Anomalist Books, 2022. The verbatim exchange about interlocking universes and television pictures is confirmed through multiple secondary sources and from the original text. The third participant in some dialogues is confirmed as psychologist Arthur Hastings.

Vallée, Jacques. Passport to Magonia: On UFOs, Folklore, and Parallel Worlds. Chicago: Henry Regnery, 1969. Reprinted Contemporary Books, 1993. First publication of the historical continuity argument and the early formulation of the control system theory. Published the same year as Blue Book's closure and the AAAS symposium.

Vallée, Jacques. The Invisible College: What a Group of Scientists Has Discovered About UFO Influences on the Human Race. New York: Dutton, 1975. The book that named the network Hynek had been building since the early 1960s. Published the same year as The Edge of Reality.

Wikipedia. Jacques Vallée. en.wikipedia.org. Confirms biographical details: born September 24, 1939, Pontoise; Sorbonne mathematics; Lille MS astrophysics; Paris Observatory; University of Texas Austin 1962; Northwestern 1963; PhD 1967 in industrial engineering and computer science. Confirms the 1978 United Nations presentation with Gordon Cooper and Hynek. Confirms Vallée as inspiration for François Truffaut's character Lacombe in Close Encounters of the Third Kind.

Amazon.com. The Edge of Reality product page. Confirms Vallée's PhD description as artificial intelligence from some sources versus industrial engineering and computer science from Wikipedia. Both descriptions refer to the same 1967 Northwestern degree. Wikipedia's more formal title is used in this text.

Wikipedia. Interdimensional hypothesis. en.wikipedia.org. Confirms Hynek as one of the hypothesis's proponents. Contains the verbatim Edge of Reality exchange: Vallée: What other wild hypotheses could we make? Hynek: There could be other universes with different quantum rules or vibration rates if you want. Our own space-time continuum could be a cross-section through a universe with many more dimensions. Think what a hard time you would have convincing an aborigine that right now, through this room, TV pictures are passing.

Sheaffer, Robert. Review: The Edge of Reality and The Invisible College. Skeptical Inquirer 1, no. 2 (Spring–Summer 1977): 74. Contains the house of cards characterization. Also documents the house of cards language and the metalogic critique of Vallée. Hynek's published response appeared in the same journal.

Clark, Jerome. The UFO Encyclopedia. 3rd ed. Detroit: Omnigraphics, 2018. Entries on Vallée, the invisible college, the interdimensional hypothesis, and the control system theory with full source bibliography.

Unidentified Phenomena. Passport to Magonia summary. unidentifiedphenomena.com. Confirms the control system theory origins in Passport to Magonia and the Charles Fort influence on Vallée's thinking about a hidden intelligence influencing human society.

Note on Vallée's degree description: Amazon and some sources describe Vallée's Northwestern PhD as being in artificial intelligence. Wikipedia describes it as industrial engineering and computer science. Both refer to the same 1967 degree. The Wikipedia description is used here as the more formally verifiable institutional title. The substance of the work — early AI research and computer database construction — is confirmed across all sources.

Chapter Eleven

Wikipedia. Center for UFO Studies. en.wikipedia.org. Confirms CUFOS founding in 1973 by Hynek. Confirms initial restriction to scientists and professionals. Confirms Allan Hendry joined in 1976. Confirms International UFO Reporter from 1976. Confirms Mark Rodeghier's succession in 1986. Confirms renaming to J. Allen Hynek Center for UFO Studies after Hynek's death. Confirms CUFOS board members including Jerome Clark, Michael Swords, and Thomas Bullard.

Enigma Labs. CUFOS: The J. Allen Hynek Center for UFO Studies. enigmalabs.io. Confirms co-founder Sherman Larsen as an insurance salesman and former NICAP member. Confirms Northwestern University's refusal to host the center. Confirms Hynek's May 1973 New Scientist interview in which he cited frustration with rambling, poorly organized, and unscientific investigation as his impetus for founding CUFOS. Confirms CUFOS's three founding tasks: investigating UAP reports, supporting research, and public dissemination.

Encyclopedia.com. J. Allen Hynek Center for UFO Studies. Quotes James W. Moseley's assessment that CUFOS was the first real attempt to set up a private research group genuinely dedicated to scientific investigations and study of UFOs. Confirms Hendry's 1979 report and its conclusion that investigators lacked adequate tools. Confirms peak years in the late 1970s and organizational transition following Hynek's move to Phoenix and subsequent death.

Disclosdex. CUFOS. disclosdex.com. Confirms UFOCAT as over 100,000 global reports. Confirms the Hynek Library in Chicago and overflow collections at Northwestern's McCormick Library. Confirms CUFOS protocols for screening cases and assigning investigators. Confirms collaboration with the Society for Scientific Exploration and data-sharing with NUFORC and MUFON.

Wikipedia. Pascagoula Abduction. en.wikipedia.org. Confirms the full chronology: October 11, 1973; Hickson age 42, Parker age 19; oval-shaped craft; three beings with claw-like hands; examination by robotic eye; sheriff's secret recording; wire service pickup; Harder's hypnosis terminated when Hickson became too frightened; polygraph results. Confirms Hickson's Dick Cavett Show appearance November 2, 1973.

Pascagoula UFO 1973 Historical Marker. hmdb.org. Contains Hynek's verbatim assessment: these men have had a very real, frightening experience. They are absolutely honest. They have had a fantastic experience. Primary source for Hynek's public statement on the case.

Openminds.tv. Remembering Charlie Hickson. Confirms Hynek's arrival representing CUFOS and Harder's arriving representing APRO. Confirms the joint statement that there was definitely something here that was not terrestrial. Provides the 17-page Keesler AFB interrogation transcript as a separately documented source.

Wikipedia. Coyne Helicopter Incident. Thinkaboutitdocs.com and Uf ologie.patrickgross.org (official Army Reserve incident report). Confirms all case details: October 18, 1973; Port Columbus departure 10:30 PM; Coyne's 19 years flying experience; approach at over 600 knots; green beam bathing cockpit; ascent from 1,700 to 3,500 feet with collective in full-down position; magnetic compass rotating; Delong family ground witnesses. Confirms National Enquirer Blue Ribbon Panel $5,000 award. Confirms Jennie Zeidman as CUFOS investigator.

Columbus Free Press. Ohio Against the Universe. columbusfreepress.co m. Confirms Hynek's own quote calling October 1973 a memorable wave containing within one week two notable classic cases — the Pascagoula fishermen and the Ohio Coyne helicopter cases — sourced to Hynek's biography by Mark O'Connell.

MNopedia. Val Johnson Incident, 1979. mnhs.org. The most authoritative source for the Val Johnson case, produced by the Minnesota Historical Society. Confirms date August 27, 1979; County Road 5 ten miles west of Stephen; 1:40 AM; CUFOS investigator Guy Westcott's initial interview; subsequent investigation by Allan Hendry. Confirms Good Morning America appearance September 11, 1979. Confirms Johnson's eventual move to Eau Claire, Wisconsin.

Wikipedia. Val Johnson Incident. en.wikipedia.org. Confirms physical damage: cracked windshield, broken headlight, damaged red emergency light, bent antenna. Confirms 39-minute unconsciousness. Confirms 14-minute discrepancy on both car clock and wristwatch. Confirms no aircraft in area per FAA and Air Force. Confirms CUFOS assessment of Johnson's account as credible.

Podcast UFO. The Case of Val Johnson. podcastufo.com. Confirms CUFOS devoted nearly the entire September-October 1979 IUR to the case, stopping the presses to include it. Contains Hendry's verbatim assessment: the biggest mystery about the Val Johnson case is trying to find one neat explanation for something that could behave the way he described yet create the kind of damage that we analyzed and discovered. Confirms the straight-forward, just-the-facts reporting style CUFOS applied.

Kare11.com. Minnesota's Most Notorious UFO Sighting. Confirms Ford crash investigator quote: I have not seen anything like this before. Confirms Sheriff Dennis Brekke's personal vouching for Johnson's honesty. Confirms Honeywell engineer Roland Wardell's conclusion that no single earthly explanation could account for all the vehicle damage, some consistent with an electrical field.

Hendry, Allan. The UFO Handbook: A Guide to Investigating, Evaluating, and Reporting UFO Sightings. Garden City: Doubleday, 1979. Hendry's comprehensive methodology report from his CUFOS investigations, confirming the 90-95 percent resolution rate and the nature of the genuine residuals.

Chapter Twelve

Britannica Blog. 30 Years of Close Encounters: Spielberg, Hynek, and UFOs. blogs.britannica.com, November 2007. The most detailed single account of the Hynek-Spielberg relationship. Confirms Spielberg's 1997 documentary quote about Hynek verbatim. Confirms cameo timing to the second: 2:02:57 into the 137-minute 1998 Collector's Edition. Confirms Hynek's objection to the Flight 19 association. Confirms the Portage County police chase as partial basis for the Dreyfuss character's encounter.

The Ankler. UFOs, Aliens and Steven Spielberg's 20-Year Obsession. t heankler.com, November 2024. Confirms Spielberg's development deal with Columbia in late 1973. Confirms Spielberg's formative Arizona meteor shower experience. Confirms the UFOs and Watergate plot concept

Spielberg considered. Confirms the Watch the Skies title origin from The Thing (1951) per McBride biography. Confirms principal photography beginning May 16, 1976.

AFI Catalog. Close Encounters of the Third Kind (1977). catalog.afi.com. The most authoritative production history source. Confirms the title history: Watch the Skies, Kingdom Come (Schrader draft), Meeting of the Minds (Hill draft), Watch the Skies again in February 1976 HR trade paper, then Close Encounters restored after Hynek's attorney's letter and Columbia settlement. Confirms filming in Mobile, Alabama aircraft hangar. Confirms Spielberg relied heavily on Hynek per March 14, 1977 LAT article.

Balaban, Bob. Spielberg, Truffaut and Me. London: Titan Books, 2002. Source for July 23, 1976 diary entry describing Hynek's arrival on the Mobile set in a Hawaiian shirt, the Van Dyck beard and Wizard of Oz comparison, the tape recorder detail, and the evening lecture to cast and crew including Dreyfuss and Dillon. Primary eyewitness source for the set visit. Originally published as Close Encounters of the Third Kind Diary.

Wikipedia. Close Encounters of the Third Kind. en.wikipedia.org. Confirms November 17, 1977 premiere at Ziegfeld Theater. Confirms over $300 million worldwide gross and third highest-grossing film of 1977. Confirms Academy Award nominations. Confirms Library of Congress National Film Registry selection December 2007. Confirms Hynek's verbatim quote about the film. Confirms the title derived from chapter 10 of The UFO Experience.

IMDB. Close Encounters of the Third Kind (1977). imdb.com. Lists Hynek's character credit as Smoking Pipe at Landing Site. Confirms uncredited appearance.

CTVA. The Dick Cavett Show, Season 8 (1973-74). ctva.biz. Confirms the November 2, 1973 episode featuring Carl Sagan, two UFO witnesses, and Captain Lawrence Coyne — the episode matching the post-Pascagoula and post-Coyne incident context in which Hynek was also active. Note: this source lists the episode as featuring Sagan and Coyne; Hynek's specific presence on this episode requires verification against the original broadcast record or CTVA expanded listings.

Biography.com. J. Allen Hynek: Project Blue Book, Quotes and Facts. Confirms Hynek as a well-known face of ufology in the 1970s through appearances on The Dick Cavett Show and In Search Of. Confirms he was a popular and well-compensated speaker on the college circuit. Confirms the 1978 United Nations presentation.

Wikipedia. J. Allen Hynek. en.wikipedia.org. Contains the Globe and Mail quote verbatim: I have looked up many times, and not seen one thing I would report. But then, I would not report a UFO without a witness. Confirms the green fireball over the desert detail. Confirms the swamp gas phrase's cultural persistence.

Wikipedia. Jacques Vallée. en.wikipedia.org. Confirms Vallée as the inspiration for François Truffaut's character Claude Lacombe in the film. This detail connects the Spielberg film back to the Hynek-Vallée intellectual partnership documented in Chapter 10.

Hynek, J. Allen. The Hynek UFO Report: What the Government Suppressed and Why. New York: Dell, 1977. Published the same year as Close Encounters, this book represents Hynek's public consolidation of his Blue Book critique in the context of the film's cultural moment.

Note on the Dick Cavett Show appearance: The November 2, 1973 episode confirmed from CTVA records features Carl Sagan described as a professor of astronomy at Northwestern — this may be a source error as Sagan was at Cornell, not Northwestern; Hynek was the Northwestern astronomer. The episode clearly covers the 1973 UFO wave. Whether Hynek was the unnamed second scientist on this specific episode, or whether his Dick Cavett appearances were on separate episodes, requires verification against complete show records or the surviving broadcast footage at Global Image Works.

Chapter Thirteen

O'Connell, Mark. The Close Encounters Man: How One Man Made the World Believe in UFOs. New York: Dey Street Books / William Morrow, 2017. The most recent and comprehensive biography of Hynek. Confirms the May 5, 1910 rooftop Halley's Comet scene, drawn from family accounts. The WTTW Chicago interview with O'Connell provides the extended passage about the rooftop and the comet.

WTTW Chicago News. How a Controversial Chicago Astronomer Influenced Close Encounters. news.wttw.com, July 17, 2017. Q&A with O'Connell, confirming the May 5 rooftop scene and the comet bookend. Source for the extended passage about Joseph and Bertha Hynek taking their five-day-old son to the roof.

Northwestern University Archives. Finding Aid: J. Allen Hynek Papers, 1910–1986. findingaids.library.northwestern.edu. Confirms retirement date of June 1978. Confirms CUFOS moved to his home in 1981 and toll-free hotline disconnected in 1982. Confirms Corralitos Observatory donation and CARA. Confirms death on April 27, 1986 at Memorial Hospital in Scottsdale. Confirms Mimi Hynek died in 1996. Primary institutional source for the personal chronology.

Franch, John. The Secret Life of J. Allen Hynek. Skeptical Inquirer 37, no. 1 (January/February 2013). The article documenting Hynek's lifelong occult interests, the Rosicrucian and Steiner connections, and the argument that his public skepticism was a façade for public consumption. Essential for understanding the full personal portrait. Draws on the Vallée journals and the Northwestern University Archives.

Vallée, Jacques. Forbidden Science: Journals 1970–1979. Volume 2. Berkeley: North Atlantic Books, 2009. Contains the accounts of Hynek's UFO Experience creating faculty antagonism at Northwestern and the deteriorating institutional relationship through the 1970s.

Vallée, Jacques. Forbidden Science: Journals 1980–1989. Volume 3. Documenting Hynek's Scottsdale move, the Jeffery Kaye backing, Vallée's meeting with the partners involved, the money-making operation remark, and Hynek's own admission that he had rushed. The November 10, 1984 entry is the primary source for the Scottsdale period. Published 2016.

Podcast UFO. Dr. J. Allen Hynek and the International Center for UFO Research. podcastufo.com. The most detailed account of the Scottsdale period available online, drawing on Keith Basterfield's blog Why did J.

Allen Hynek move to Scottsdale? (October 8, 2017) and the Vallée journals. Confirms the Jeffery Kaye backing details and the disintegration of the partnership.

Coast to Coast AM. The Life of J. Allen Hynek. George Noory host, Paola Harris and Ann Eller guests. December 10, 2009. concen.org transcript. Source for Ann Eller's portrait of Hynek in his final year: very quiet, private, soft-spoken man with a twinkle in his eye. Source for Eller's description of him as troubled by the esoteric aspects of the phenomenon. Source for the closet Rosicrucian characterization from Eller. Source for Harris's statement that she cried her eyes out when he died.

Chicago Tribune. J. Allen Hynek, 75, NU Professor and International Expert on UFOs. April 30, 1986. Primary source obituary. Contains the verbatim passage from the 1977 book introduction: We know so little about the vast universe. Services confirmed as private.

Washington Post. Dr. J. Allen Hynek, 75, a leading authority on unidentified flying objects. May 3, 1986. Contemporaneous obituary confirming cause of death and basic biographical details.

Wikitree. Josef Allen Hynek (1910–1986). wikitree.com. Confirms first marriage December 24, 1932 in Washington County, Arkansas to Martha Doone Alexander. Confirms second marriage May 31, 1942 in Medina, Ohio to Miriam Curtis.

Wikipedia. J. Allen Hynek. en.wikipedia.org. Confirms five children with Mimi. Confirms Joel Hynek as Oscar-winning visual effects supervisor. Confirms Hynek's statement to Vallée about becoming an astronomer to discover the very limitations of science, the places where it broke down.

The Vallée source cited is Forbidden Science volume 2 page 232, 1996 Marlowe & Company edition.

Ridpath, Ian. Interview with J. Allen Hynek. New Scientist, May 17, 1973. Source for Hynek's statement: I've never launched any new theories, I've never made any outstanding discoveries. Source for Ridpath's response that Hynek would be remembered as the man who made UFOs respectable. Cited in the encyclopedia.com entry on Hynek.

Hynek, J. Allen, Philip Imbrogno, and Bob Pratt. Night Siege: The Hudson Valley UFO Sightings. New York: Ballantine Books, 1987. Hynek's final book, published posthumously. Documents the 1983–1985 Hudson Valley sighting wave, Hynek's last major case investigation.

Biography.com. J. Allen Hynek. Confirms born May 1, 1910, the spring Halley's Comet passed close to Earth; died April 27, 1986, shortly after the comet's return appearance. This biographical coincidence is noted in the biography.com article as a notable fact.

Note on Hynek's place of burial: The Find A Grave entry lists Green Acres Memorial Park in Scottsdale but notes that the cemetery office has no record of this person buried or cremated there. His actual burial location is unconfirmed online and should be verified through the family, CUFOS, or Arizona vital records before the final manuscript.

Chapter Fourteen

Hynek, J. Allen. Kepler and the SKY. Sky and Telescope 50, no. 6 (December 1975): 451–455. Primary source for the Kepler essay passage about a tenuous bond linking present-day astronomy's metaphysical thinking and

Kepler's brand of metaphysics, and the characterization of both systems as the repository of fundamental questions not entertained on the present playing field of physical science. The essay date, journal, and page numbers are confirmed from the Franch Skeptical Inquirer article citation.

Franch, John. The Secret Life of J. Allen Hynek. Skeptical Inquirer 37, no. 1 (January/February 2013). Available at skepticalinquirer.org. The primary source for Hynek's three documented late-life formulations: observational devices materialized into our world by the denizens of another (1967, from Vallée Forbidden Science vol. 1, page 306); psychic projections created by an extradimensional intelligence in some parallel reality (Gardner 1997, page 253); and elementals — nature spirits — as told to Jerome Clark (Clark 1998). Also primary source for the Kepler essay passage and Hynek's statement about modern astronomy as a playing field for the metaphysicist. The wishful thinking verdict and the quote about UFOs as the beyond where science could not reach are from this article.

Halperin, David. Truth with Meaning: The Quest of J. Allen Hynek. davidhalperin.net, January 25, 2021. The most thoughtful counterreading of the Franch article. Argues that the mystical tilt and the scientific rigor are related rather than contradictory phenomena, and that Hynek's restless searching independent mind is more accurately described as the source of his achievement than as evidence of its disqualification. The passage about whether Hynek would have been capable of the same investigation without the prior attraction to the territory beyond science is from this essay.

Vallée, Jacques. Forbidden Science: Journals 1957–1969. Berkeley: North Atlantic Books, 1992. Page 232: Hynek's statement to Vallée that he had become an astronomer to discover the very limitations of science, the places where it broke down, the phenomena it didn't explain. Page 306: Hynek's 1967 speculation about observational devices materialized into our world by the denizens of another. Also contains the Colorado gesture toward the landscape with I have never stopped thinking about what must lie beyond all this.

Hynek, J. Allen. The Hynek UFO Report: What the Government Suppressed and Why. New York: Dell, 1977. Available in full text at archive.o rg. Primary source for the 1977 formulation on abduction cases: CE3 cases hold the most fascination because they bring into focus the concept of other intelligences in space and the possibility of intelligent contact with all that it might imply. Also primary source for the methodological discussion of regressive hypnosis as producing data of a different evidential status than waking reports.

Hynek, J. Allen, and Jacques Vallée. The Edge of Reality: A Progress Report on Unidentified Flying Objects. Chicago: Henry Regnery, 1975. Primary source for the interlocking universes exchange and the formulation that the UFO represents an unknown but real phenomenon whose implications take us to the very edge of what we consider the known and real in our physical environment.

Gardner, Martin. Ufo's and Alien Beings. In Weird Water and Fuzzy Logic: More Notes of a Fringe Watcher. Amherst: Prometheus Books, 1996. Cited in Franch as Gardner 1997, page 253, for the psychic projections formulation. The date discrepancy between Franch's 1997 citation and the 1996 publication date may reflect a different edition or reprint. The passage is attributed to Hynek in both the Franch article and the Clark UFO Encyclopedia entry.

Clark, Jerome. The UFO Encyclopedia. 2nd ed. Detroit: Omnigraphics, 1998. Contains the account of Hynek's conversation with Clark in which he described elementals — nature spirits — as possibly behind the UFO phenomenon. This is Hynek's most radical late-life formulation and is documented by Clark as a direct conversation rather than a published statement. The third edition (2018) presumably preserves this account.

Vallée, Jacques. Passport to Magonia: On UFOs, Folklore, and Parallel Worlds. Chicago: Henry Regnery, 1969. Primary source for the evasiveness argument — the phenomenon's tendency to be exactly as convincing as needed to produce a specific effect without being convincing enough to constitute definitive proof. The theoretical foundation for Hynek's late-life uncertainty about the relationship between observation and the phenomenon itself.

Wikipedia. Interdimensional UFO hypothesis. en.wikipedia.org. Documents the full intellectual lineage of the hypothesis from Meade Layne through Keel, Hynek, and Vallée. Contains the Kripal quotation from Rice University about the long life of the interdimensional reading in ufology. Confirms Hynek as one of the hypothesis's principal scientific proponents.

Hynek, J. Allen. Unusual Aerial Phenomena. Journal of the Optical Society of America 43, no. 4 (April 1953): 311–314. The foundational statement of the philosophy that drives the entire biography: Ridicule is not part of the scientific method, and people should not be taught that it is. Placed here as the philosophical anchor of Chapter 14 because it is the earliest available formulation of the position that defines Hynek's intellectual life.

Note on the elementals claim: The account of Hynek telling Clark that elementals were behind the UFO phenomenon is documented in Clark's UFO Encyclopedia and reproduced in the Franch article. It is the most radical of Hynek's documented late-life positions and should be presented as one point in a range of evolving formulations rather than as a fixed conclusion. The evidence that these were all provisional hypotheses rather than adopted beliefs is Hynek's own consistent insistence on the distinction between evidence and conclusion throughout his published work.

Epilogue

Cooper, Helene, Leslie Kean, and Ralph Blumenthal. Glowing Auras and Black Money: The Pentagon's Mysterious U.F.O. Program. New York Times, December 16, 2017. The foundational article of the current UAP era. Confirms AATIP, Harry Reid's $22 million funding, Elizondo's role, and the three Navy videos.

United States Congress. House Committee on Oversight and Accountability, Subcommittee on National Security, the Border, and Foreign Affairs. Unidentified Anomalous Phenomena: Exposing the Truth. Hearing, 118th Congress, July 26, 2023. Primary source transcript available at congress.gov. Contains Grusch's sworn statement, verbatim passage I like to use the term non-human, I don't like to denote origin, it keeps the aperture open scientifically, and Burchett's direct questions regarding evidence of harm to those attempting disclosure.

United States Congress. House Intelligence Subcommittee on Counterterrorism, Counterintelligence, and Counterproliferation. Hearing on Unidentified Aerial Phenomena. 117th Congress, May 17, 2022. First public congressional hearing on the subject in over 50 years, confirmed across all UAP timeline sources. Featured testimony from Pentagon officials Ronald Moultrie and Scott Bray.

National Defense Authorization Act for Fiscal Year 2022. Public Law 117-81. Contains the Gillibrand-Rubio amendment establishing AARO. Available at congress.gov. Confirms the transition from UAPTF to AARO and the expanded transmedium mandate.

All-domain Anomaly Resolution Office. Historical Record Report on the U.S. Government's Involvement with Unidentified Anomalous Phenomena. Washington: Department of Defense, March 2024. The AARO report finding no evidence of extraterrestrial spacecraft after reviewing nearly eight decades of cases. Disputed by multiple whistleblowers. Sean Kirkpatrick's directorship ended December 2023.

National Aeronautics and Space Administration. UAP Independent Study Team Report. Washington: NASA, September 2023. The first formal engagement with the subject by a major government science agency since the Condon Report. Established a Director of UAP Research within NASA. Acknowledged the need for standardized data collection and rigorous methodology — the same need Hynek identified fifty years earlier.

Kirkpatrick, Sean. Quoted in Luscombe, Richard. It Is Basically a Religion: Skeptical Ex-Pentagon UFO Chief Blasts Conspiracy Theorists. The Guardian, December 2023. Kirkpatrick's characterization of UAP advocacy as a religious belief that transcends critical thinking and rational thought. The precise echo of Robertson Panel era dismissal that Hynek spent his career arguing against.

Hynek, J. Allen. Axioms for a Developing Science of UFOs. Paper presented at the Joint Symposium of the American Institute of Aeronautics and Astronautics, Los Angeles, 1975. Primary source for the list of phenomena requiring explanation beyond present theoretical capabilities, including materialization and dematerialization, EM effects, physiological effects, and structured occupant reports.

CUFOS. J. Allen Hynek Center for UFO Studies. cufos.org. Confirms Mark Rodeghier's continued leadership since 1986. Confirms ongoing case archive maintenance. Confirms UFOCAT database preservation. Confirms CUFOS materials cited in NASA's 2023 UAP study panel background research.

New Space Economy. Congressional Oversight and the Evolution of UAP Transparency. newspaceeconomy.ca, November 2025. Comprehensive timeline of the congressional engagement from 2017 through 2025 NDAA provisions. Confirms the November 2024 hearing as the third major public hearing. Confirms the Schumer-Rounds Disclosure Act provisions and their subsequent stripping from the final NDAA.

Bolender, Carroll H. Memorandum: Unidentified Flying Objects. October 20, 1969. Declassified through FOIA 1979. Available through National Archives. The document confirming that cases capable of affecting national security were not part of the Blue Book system and were handled through other channels. The fifty-year institutional continuity between this document and the AATIP revelations of 2017 is the Epilogue's central historical argument.

Hynek, J. Allen. Unusual Aerial Phenomena. Journal of the Optical Society of America 43, no. 4 (April 1953). Ridicule is not part of the scientific method, and people should not be taught that it is. The foundational statement that defines the entire intellectual life documented in this biography, and whose continued relevance the current UAP moment demonstrates.

AppendIx A Complete Hynek Bibliography

This bibliography lists the published works of J. Allen Hynek in chronological order, divided by category. It draws on the Northwestern University Archives finding aid, the CUFOS bibliographic records, the O'Connell biography, and the Franch Skeptical Inquirer article. Items marked [VERIFY] require confirmation against the Northwestern University Archives or the CUFOS collection before final publication. Items marked [POSTHUMOUS] were published after Hynek's death on April 27, 1986.

I. Books — Scientific (Astronomy)

Hynek, J. Allen, ed. Astrophysics: A Topical Symposium Commemorating the Fiftieth Anniversary of the Yerkes Observatory and a Half Century of Progress in Astrophysics. New York: McGraw-Hill, 1951. Edited volume drawing on the 1947 Yerkes Observatory jubilee symposium. Hynek's most significant purely scientific publication as an editor. Copy held in the Northwestern University Archives Faculty Authors Collection.

Hynek, J. Allen, and Norman D. Anderson. Challenge of the Universe. Washington, D.C.: National Science Teachers Association, 1962. A Vistas of Science book. Co-authored with Anderson; illustrated by Helmut Wimmer. Popular science education volume aimed at secondary school audiences. Copy held in the Northwestern University Archives Faculty Authors Collection.

Hynek, J. Allen, and Necia H. Apfel. Astronomy One. Menlo Park, California: W. A. Benjamin, 1972. Introductory university astronomy textbook, co-authored with Apfel. Copy held in the Northwestern University Archives Faculty Authors Collection.

II. Books — UFO Investigation

Hynek, J. Allen. The UFO Experience: A Scientific Inquiry. Chicago: Henry Regnery, 1972. Reprinted New York: Ballantine Books, 1974. Reprinted Cambridge, Massachusetts: Da Capo Press, 1998. ISBN 978-1-56924-782-2. Hynek's foundational work. Introduces the six-category close encounter classification system. Dedicated to the witnesses. Reviewed in Science 177 (1972) by Bruce C. Murray. The classification system used in this book is drawn directly from chapter 10. Copy held in the Northwestern University Archives Faculty Authors Collection.

Hynek, J. Allen, and Jacques Vallée. The Edge of Reality: A Progress Report on Unidentified Flying Objects. Chicago: Henry Regnery, 1975. Reprinted San Antonio: Anomalist Books, 2022. ISBN 978-1-949501-17-8. Structured as dialogue between Hynek and Vallée, with participation by psychologist Arthur Hastings in selected sections. Introduces the interdimensional hypothesis in extended form.

Hynek, J. Allen. The Hynek UFO Report: What the Government Suppressed and Why. New York: Dell, 1977. Revised and reissued as The Hynek UFO Report: The Authoritative Account of the Project Blue Book Cover-Up. San Antonio: Anomalist Books, 2020. ISBN

978-1-59003-303-6. Forewords by Paul Hynek, Scott Hynek, Roxane Hynek, Joel Hynek, and Ross Hynek. Full text available at archive.org. Hynek's retrospective assessment of Blue Book with case analyses.

Hynek, J. Allen, Philip Imbrogno, and Bob Pratt. Night Siege: The Hudson Valley UFO Sightings. New York: Ballantine Books, 1987. [POSTHUMOUS] Hynek's final book, documenting the 1983–1985 Hudson Valley, New York sighting wave. Co-authored with Imbrogno and Pratt; Hynek died April 27, 1986, before the book's completion and publication.

III. Peer-Reviewed Journal Articles

Hynek, J. Allen. Unusual Aerial Phenomena. Journal of the Optical Society of America 43, no. 4 (April 1953): 311–314. The most significant of Hynek's journal publications on the UFO subject and the source of his best-known statement: Ridicule is not part of the scientific method, and people should not be taught that it is. Published while Hynek was still serving as Blue Book's scientific consultant.

Hynek, J. Allen. The Condon Report and UFOs. Bulletin of the Atomic Scientists 25, no. 4 (April 1969): 39–42. Hynek's formal peer-reviewed critique of the Condon Report, published the same month the Colorado study was made public. Argues that the body of the report constitutes as good an argument for continued study as could have been made in a short time. Available through the Bulletin of the Atomic Scientists archive.

Hynek, J. Allen. Twenty-One Years of UFO Reports. In UFOs: A Scientific Debate, edited by Carl Sagan and Thornton Page, 37–51. Ithaca: Cornell University Press, 1972. Originally presented at the 134th Meeting of the American Association for the Advancement of Science, Boston, December 27, 1969. First public presentation of the six-category taxonomy in a formal scientific context.

Hynek, J. Allen. Kepler and the SKY. Sky and Telescope 50, no. 6 (December 1975): 451–455. Essay on Johannes Kepler's metaphysical cosmology, simultaneously a defense of Hynek's own late-career philosophical position. Published the same year as The Edge of Reality.

Hynek, J. Allen. Axioms for a Developing Science of UFOs. Paper presented at the Joint Symposium of the American Institute of Aeronautics and Astronautics and the American Astronautical Society, Los Angeles, 1975. Published in the symposium proceedings. Contains Hynek's clearest late-career statement of phenomena requiring explanation beyond present theoretical capabilities. [VERIFY: proceedings publisher and page numbers against CUFOS archives.]

IV. Major Popular Articles

Hynek, J. Allen. Are Flying Saucers Real? Saturday Evening Post, December 17, 1966. Hynek's first major public break with the institutional Blue Book position, published under his own name in a mass circulation magazine. Available through the Ann Druffel Special Collection of the Expanding Frontiers Archive and through ProQuest Historical Newspapers.

Hynek, J. Allen. The UFO Gap. Playboy, December 1967. Extended popular argument for serious scientific investigation. Contains the statement that a proper investigation would produce, within a year, photographs, movies, spectrograms, plaster casts of indentations, and detailed measurements. Available through the Expanding Frontiers Archive.

Hynek, J. Allen. Matters Celestial. Science Digest, monthly column beginning July 1982. Monthly astronomy and science column for a general audience, running from at least July 1982. The July 1982 inaugural column is documented in the AbeBooks catalog entry for that issue. [VERIFY: full run dates and column count against Science Digest archive.]

V. Congressional Testimony

Hynek, J. Allen. Statement before the House Committee on Armed Services. 89th Congress, 2nd Session. April 5, 1966. Available through the National Archives and the Internet Archive (ia600300.us.archive.org) . Hynek's first congressional appearance calling for a civilian panel of physical and social scientists to study UFOs — the first time he publicly contradicted the institutional Blue Book position in a government context.

Hynek, J. Allen. Testimony before the House Committee on Science and Astronautics, Symposium on Unidentified Flying Objects. 90th Congress, 2nd Session. July 29, 1968. Full transcript published as: United States Congress. House Committee on Science and Astronautics. Symposium on Unidentified Flying Objects: Hearings. Washington: Government Printing Office, 1968. Available through NICAP archives (nicap.org) and the Government Publishing Office. Hynek's first congressional appearance as a private scientist rather than an Air Force representative.

VI. International UFO Reporter — Selected Contributions

Hynek, J. Allen. International UFO Reporter, various issues, 1976–1985. As founder of CUFOS, Hynek contributed editorials, case analyses, and commentary to the International UFO Reporter from its founding in 1976 through approximately 1985. A complete run of the IUR is held in the CUFOS archives, Chicago, and selected issues are available through the CUFOS website (cufos.org). The November-December 1977 issue contained Hynek's report on the Hudson Valley wave. [VERIFY: complete list of Hynek-authored pieces in IUR against CUFOS archive holdings.]

VII. Forewords, Introductions, and Contributed Chapters

Hynek, J. Allen. Foreword to Clear Intent: The Government Cover-Up of the UFO Experience, by Lawrence Fawcett and Barry J. Greenwood. Englewood Cliffs, New Jersey: Prentice-Hall, 1984. Documented in the AbeBooks catalog entry for the first edition.

Hynek, J. Allen. Various contributions to the Journal of UFO Studies. CUFOS, 1979–1985. Hynek contributed to the Journal of UFO Studies, CUFOS's peer-reviewed publication, through its early years. [VERIFY: specific article titles, volumes, and dates against CUFOS archive holdings.]

VIII. Significant Interviews in the Public Record

Ridpath, Ian. Interview with J. Allen Hynek. New Scientist, May 17, 1973. Contains Hynek's statement that he had never launched any new theories nor made outstanding discoveries, and Ridpath's assessment that Hynek would be remembered as the man who made UFOs respectable.

Hynek, J. Allen. Interview. Omni Magazine, 1985. Contains Hynek's statement: I was a thorough skeptic, and I'm afraid I helped to engender the idea that it must be nonsense, therefore it is nonsense. Also cited in Newsweek, January 8, 2019. [VERIFY: exact issue date and page numbers against Omni archive.]

Hynek, J. Allen. Interview. Globe and Mail (Canada), date unrecorded. Contains the statement: I have looked up many times and not seen one thing I would report. But then, I would not report a UFO without a witness. I have seen some strange things — I saw a green fireball over the desert a few years ago — but I would not report it without a witness. [VERIFY: date and page number against Globe and Mail archive.]

Hynek, J. Allen. Interview. Fate Magazine, various dates. Hynek contributed to and was interviewed by Fate Magazine across his independent investigation years. [VERIFY: specific interview dates and issues against Fate archive.]

IX. Archival Holdings — Primary Sources

J. Allen Hynek Papers, 1910–1986. Northwestern University Archives, Northwestern University Library, Evanston, Illinois. Finding aid at findingaids.library.northwestern.edu/repositories/6/resources/373. Accessions 74-44, 79-140, 82-138, and 88-269. Contains biographical materials, correspondence, newspaper and journal clippings 1950s–1988 arranged chronologically, teaching notes, Blue Book related materials, Project Stargazer reports, photographs and negatives of reported UFOs, Corralitos and Organ Pass Observatory materials. Note: the finding aid states the collection offers scant insight into personal affairs and is missing personal correspondence and private papers.

CUFOS Archives. J. Allen Hynek Center for UFO Studies, Chicago, Illinois. cufos.org. Contains Hynek's CUFOS-period correspondence, case files, UFOCAT database records, International UFO Reporter complete run, Journal of UFO Studies complete run, and unpublished manuscripts.

Contact Mark Rodeghier, Scientific Director, for access to specific holdings.

Expanding Frontiers Archive. Ann Druffel Special Collection. Contains Hynek's Saturday Evening Post article (1966), Playboy article (1967), and other primary popular press contributions. Access through the archive's cataloging system.

Note on completeness: This bibliography draws on verified published sources and archival finding aids. The complete record of Hynek's journal articles, conference papers, and minor popular contributions has not been exhaustively compiled in any single publicly available source. A definitive bibliography requires systematic examination of the Northwestern University Archives and the CUFOS holdings. Items marked [VERIFY] represent the highest-priority gaps for the final manuscript.

Appendix B Chronology of Major Cases

This chronology covers the cases most significant to the narrative of J. Allen Hynek's investigation, from the inception of Project Sign through the final CUFOS investigations. Each entry includes date, location, witness category, Hynek's role, Blue Book or CUFOS disposition, and primary source citations. Cases are cross-referenced to the chapters in which they are discussed. The term Hynek's role distinguishes between cases he investigated directly in the field, cases he reviewed analytically, and cases investigated by CUFOS under his directorship. Items marked [VERIFY] require primary source confirmation before the final manuscript.

Project Sign Era — 1947–1948

Mantell Incident. January 7, 1948. Godman Army Airfield, Fort Knox, Kentucky. Witnesses: Captain Thomas F. Mantell, Kentucky Air National Guard, and tower personnel at Godman AFB. Object described as circular, 250 to 300 feet in diameter, moving westward; Mantell described it in his last radio transmission as metallic and of tremendous size. Mantell pursued the object to approximately 22,500 feet without supplemental oxygen, lost consciousness, and crashed near Franklin, Kentucky. His wristwatch stopped at 3:18 p.m. Hynek's role: reviewed analytically for Project Sign.

Initial Hynek explanation: Venus. Hynek retracted this explanation in 1952, stating Venus was not bright enough to be seen and considerable haze was present. Final Blue Book disposition: probable Skyhook balloon, a classified Navy program at the time of the incident. The Mantell case, along with Chiles-Whitted and the Gorman Dogfight, was one of the three classic 1948 cases that convinced Project Sign personnel that UFOs represented a genuine physical phenomenon. Chapter reference: Chapter Three. Sources: Wikipedia, Mantell UFO incident; Completely Kentucky Wiki; This Day in Aviation (thisdayinaviation.com); Project Blue Book case file, National Archives.

Chiles-Whitted Encounter. July 24, 1948. Approximately 20 miles southwest of Montgomery, Alabama. Witnesses: Captain Clarence S. Chiles and First Officer John B. Whitted, Eastern Airlines DC-3, and passenger Clarence L. McKelvie. Object described as torpedo-shaped, approximately 100 feet long, double row of windows emitting bluish-white light, no wings, moving at extraordinary speed. Near-collision at 5,000 feet, 2:45 a.m. Ground corroboration: military observer at Robins AFB, Georgia, reported identical object one hour earlier. Hynek's role: reviewed analytically for Project Sign. Hynek explanation: bright meteor. Disposition: core evidence for Estimate of the Situation (see below). The case remained in the literature as one of the most credible early pilot sightings and the primary catalyst for the Estimate. Chapter reference: Chapter Three. Sources: Wikipedia, Chiles-Whitted UFO encounter; History.co m; Grokipedia; Biblioteca Pleyades, Project Sign analysis.

Gorman Dogfight. October 1, 1948. Fargo, North Dakota. Witness: Second Lieutenant George F. Gorman, North Dakota Air National Guard,

F-51D Mustang. Additional witnesses: another pilot, a passenger, and tower operator Lloyd D. Jensen at Hector Field. Twenty-seven-minute aerial encounter with a white luminous object demonstrating superior maneuverability and apparent evasive intelligence. Near head-on collision at 500 feet. Project Sign tested Gorman's aircraft for radioactivity, finding no anomalous levels. Hynek's role: reviewed analytically for Project Sign. Disposition: one of three 1948 classic cases for Project Sign; Blue Book later listed as possible weather balloon or atmospheric phenomenon. Chapter reference: Chapter Three. Sources: Black Box UFO (blackboxufo.com); The Big Study blog; Wikipedia, Gorman Dogfight.

Estimate of the Situation. September–October 1948. Air Technical Intelligence Center, Wright Field, Ohio. Document: classified Top Secret assessment authored by Project Sign director Captain Robert R. Sneider and personnel. Conclusion: UFOs are of interplanetary origin. Core evidentiary case: Chiles-Whitted, supported by Gorman Dogfight and multiple earlier 1947–1948 sightings. Chain of command: approved by Colonels Clingerman and McCoy, submitted to General Charles Cabell, forwarded to General Hoyt S. Vandenberg, Air Force Chief of Staff. Vandenberg's disposition: rejected October 1948 on grounds that the evidence was insufficient to support its conclusions. Document copies ordered destroyed; no surviving copy has been located through FOIA requests as of this writing. Ruppelt stated he personally read a copy before it was destroyed. Hynek's role: consulting astronomer for Project Sign during this period. Chapter reference: Chapters Three and Four. Sources: Military History Fandom Wiki, Estimate of the Situation; HowStuffWorks, Project Sign; Biblioteca Pleyades, Project Sign; The Big Study blog.

Project Blue Book Era — 1952–1969

Robertson Panel. January 14–17, 1953. Central Intelligence Agency, Washington, D.C. Convened by: CIA Scientific Intelligence Division. Chair: H. P. Robertson, California Institute of Technology. Panel members: Luis Alvarez (UC Berkeley), Lloyd Berkner (Brookhaven National Laboratory), Samuel Goudsmit (Brookhaven), and Thornton Page (Johns Hopkins). Hynek's role: associate member, attended proceedings, pushed back with Page against the dismissive tone of some panel members. Conclusion: UFOs do not directly threaten national security, but public interest in UFOs does constitute a national security threat; recommended active debunking program and reduction of public reporting. Classified until 1966. Hynek's private assessment: the panel reached conclusions predetermined by its institutional mandate. Chapter reference: Chapter Four. Sources: Wikipedia, Robertson Panel; Clark, UFO Encyclopedia (2018).

Socorro Landing. April 24, 1964. Socorro, New Mexico. Witness: Officer Lonnie Zamora, Socorro Police Department. Object: egg-shaped, on four landing legs, two figures in white coveralls with red insignia. Physical trace evidence: four landing pad impressions in the soil, four burned and smoldering patches of grass or brush. Hynek investigated the site in person on two occasions: immediately after the incident and on August 15, 1964. Hynek's letter to Donald Menzel, September 1964: called Zamora potentially the Rosetta Stone of UFO investigation and the most unimpeachable witness in sixteen years of Blue Book. Disposition: Blue Book listed Unknown. Major Quintanilla's concurrent assessment: one of the best documented cases on record; declined to pursue. Blue Book case number: 8766. Chapter reference: Chapter Five. Sources: Wikipedia, Lonnie Zamora incident; NICAP Socorro case file; Hynek letter to Menzel, September 1964, cited in O'Connell (2017).

Michigan Dexter-Hillsdale Sightings. March 14–21, 1966. Lima Township and Hillsdale, Michigan. Primary witnesses: Frank Mannor and son (Dexter), eighty-seven witnesses at MacIntyre Residence Hall, Hillsdale College, including William Van Horn, Hillsdale County civil defense director. Duration of Hillsdale observation: approximately 9 p.m. to 5:10 a.m. Hynek arrived Dexter area March 23, 1966. Press conference: Detroit Press Club, March 25, 1966. Hynek explanation issued: marsh gas. Sheriff Douglas Harvey's account: Hynek told him privately before the press conference that there was something here that could not be identified; made one phone call to Washington and emerged with the marsh gas explanation. Blue Book disposition: explained as marsh gas and astronomical phenomena. Congressional response: Gerald Ford letter to House Armed Services and Science Committees, March 28, 1966. Congressional hearing: House Armed Services Committee, April 5, 1966. Chapter reference: Chapters Six and Eight. Sources: Wikipedia, Michigan swamp gas; Bentley Historical Library, University of Michigan; ClickOnDetroit; Michigan Today; Black Vault.

Portage County UFO Chase. April 17, 1966. Portage County, Ohio, into Pennsylvania. Witnesses: multiple sheriff's deputies in four patrol cars, including Deputy Dale Spaur and Deputy Wilbur Neff, who pursued the object from Portage County, Ohio, through Atwater, Randolph, and into Pennsylvania, a total of approximately eighty-five miles, before the object departed upward. Duration: approximately ninety minutes. Object described as bright, round or dome-shaped, estimated twenty to forty feet wide. Additional law enforcement witnesses confirmed in Pennsylvania. Hynek's investigation: conducted personal interviews. Blue Book disposition: identified as Echo satellite and Venus — a conclusion the witnesses publicly and specifically rejected. Hynek's private assessment: the explanation was, in his own words, a classic example of forcing a case into an explanation. Chapter reference: Chapter Six (contextual). Sources: Wikipedia, Portage County UFO Chase; NICAP case file.

Condon Committee and Low Memo. August 9, 1966–November 1968. University of Colorado, Boulder. Robert Low memorandum August 9, 1966: pre-investigation internal document asserting the study would appear objective while maintaining near-zero expectation of finding a saucer. Committee director: Dr. Edward U. Condon. Project coordinator: Robert Low. Hynek briefed the committee in Boulder. Saunders and Levine fired February 1968. Low memo published in Look, May 14, 1968. NICAP severed cooperation May 1968. Condon Report released November 1968 as Scientific Study of Unidentified Flying Objects. National Academy of Sciences endorsement: January 1969. Hynek's published critique: Bulletin of the Atomic Scientists 25, no. 4 (April 1969). Chapter reference: Chapter Seven. Sources: Wikipedia, Condon Committee; Fuller, Look (May 1968); Saunders and Harkins (1968).

Project Blue Book Closure. December 17, 1969. Pentagon, Washington, D.C. Secretary of the Air Force announcement: December 17, 1969. Duration: 1947–1969. Total cases: 12,618 sightings reported. Classified Unknown: 701. Bolender Memorandum: October 20, 1969, signed by Brigadier General Carroll Bolender, recommending closure; contains key passage that reports capable of affecting national security were not part of the Blue Book system and were handled through other channels. Bolender memo declassified through FOIA, January 1979, obtained by Robert Todd. Chapter reference: Chapter Seven. Sources: Air Force press release December 17, 1969; National Archives, Project Blue Book Records, Record Group 341; Bolender memo, National Archives.

CUFOS Era — 1973–1985

Pascagoula Abduction. October 11, 1973. West bank of Pascagoula River, Pascagoula, Mississippi. Witnesses: Charles Hickson, age 42, shipyard foreman; Calvin Parker, age 19 [VERIFY against Keesler AFB interrogation transcript], welder. Object: oval-shaped, glowing, approximately 30 feet wide, 8 to 10 feet high, blue light. Three beings: approximately 5 feet tall, gray-skinned, claw-like hands, no visible eyes. Duration of encounter and examination: approximately 20 minutes. Sheriff's department secret recording: confirmed genuine distress, no indication of hoax. James Harder (APRO) arrived first; conducted hypnotic regression, terminated when Hickson became too frightened to continue. Hynek arrived representing CUFOS; conducted independent interviews. Hickson polygraph: passed. Keesler AFB medical examination: no radiation exposure. Hynek's public assessment: these men have had a very real, frightening experience; they are absolutely honest. Blue Book was closed; no official investigation. CUFOS assessment: significant unresolved case. Chapter reference: Chapter Eleven. Sources: Wikipedia, Pascagoula Abduction; historical marker HMDB.org; Openminds.tv; Hinds Community College LibGuide.

Coyne Helicopter Incident. October 18, 1973. Approximately 10 miles south of Mansfield, Ohio. Witnesses: Captain Lawrence J. Coyne (commander, 19 years flying experience), Sergeant Robert Yanacsek, Sergeant John Healey, Specialist Five Robert Yanacek; UH-1H Huey helicopter, Army Reserve. Ground witnesses: Emma Delong and four children, Route 430, independently confirmed timing and sequence. Object: gray metallic cigar-shape approximately 60 feet long and 20 feet high, red light forward, white light aft, green beam sweeping forward. Critical anomaly: helicopter rose from 1,700 to 3,500 feet with collective in full-down position, no g-forces experienced. Magnetic compass rotated; radio failed during encounter and resumed afterward. CUFOS investigator: Jennie Zeidman, former Ohio State astronomy student, former Blue Book associate. National Enquirer Blue Ribbon Panel award: $5,000 for most scientifically valuable UFO report of 1973. Chapter reference: Chapter Eleven. Sources: Wikipedia, Coyne UFO Incident; NICAP case file; Army Reserve official incident report; Columbus Free Press.

Val Johnson Incident. August 27, 1979. County Road 5, approximately 11 miles west of Stephen, Marshall County, Minnesota. Witness: Deputy Sheriff Val Johnson, Marshall County Sheriff's Department. Time: approximately 1:40 a.m. Object: intense light, rushed at the patrol car instantaneously from approximately one mile. Duration of unconsciousness: 39 minutes. Vehicle damage: cracked windshield broken from outside, shattered headlight, damaged red emergency light, bent antenna. Skid marks: 854 feet. Johnson's injuries: eye injuries consistent with welder's burns. Clock anomaly: both patrol car clock and Johnson's wristwatch running 14 minutes slow; both had been synchronized before the incident; discrepancy identical on both devices. No aircraft in area confirmed by FAA and Air Force. CUFOS first responder: Guy Westcott; primary investigator: Allan Hendry. Ford Motor Company crash examiner: stated damage pattern was unprecedented. Honeywell engineer Roland Wardell: no single earthly explanation could account for all the damage, some consistent with an electrical field. CUFOS IUR coverage: nearly entire September-October 1979 issue. Johnson's patrol car now on permanent display at Marshall County Historical Society Museum, Warren, Minnesota. Chapter reference: Chapter Eleven. Sources: MNopedia (Minnesota Historical Society); Wikipedia, Val Johnson Incident; Podcast UFO; Kare11.com.

Hudson Valley Wave. 1983–1985. Hudson Valley, New York; Westchester, Dutchess, and Putnam Counties, New York; and Fairfield County, Connecticut. Witnesses: thousands across multiple years. Object described consistently as large boomerang or chevron-shaped, hundreds of feet wide, moving slowly and silently with multicolored lights. Hynek's role: primary investigator for CUFOS; the investigation constituted his final major field work. Book: Night Siege: The Hudson Valley UFO Sightings, co-authored with Philip Imbrogno and Bob Pratt (Ballantine Books, 1987), published posthumously. CUFOS disposition: unexplained. Chapter reference: Chapter Thirteen. Sources: Hynek, Imbrogno, and Pratt (1987); Wikipedia, Hudson Valley UFO sightings.

Supplementary Cases — Referenced in Chapters but Not Primary Narrative Cases

Levelland, Texas Sightings. November 2–3, 1957. Levelland, Texas. Multiple witnesses reported a glowing egg-shaped object causing automobile engine failures and headlight extinctions across a wide area over several hours. Considered one of the most significant EM effects cases in the Blue Book files. Blue Book disposition: ball lightning, despite the absence of thunderstorm conditions. Hynek's role: reviewed analytically. Chapter reference: Chapter Four (contextual). Sources: Wikipedia, Levelland UFO case; Clark, UFO Encyclopedia (2018).

Portage County Jaroslaw Photos. January 1967. Livingston County, Michigan. Witnesses: Grant Jaroslaw and brother Dan. Polaroid photographs of alleged hamburger-shaped object. Hynek's role: assessed photographs without access to originals; offered hedged assessment that he could not identify them as a hoax. National headlines: Expert Sees No Hoax in Boys' UFO Photos. Confession: nine years later, brothers confirmed photographs were fabricated. Significance: Hynek's most instructive photographic methodology failure; thereafter he required original negatives before any photographic assessment. Chapter reference: Chapter Nine. Sources: Collins, Curt. Blue Blurry Lines.

Tehran F-4 Intercept. September 18–19, 1976. Tehran, Iran. Witnesses: two Iranian Air Force F-4 Phantom crews; civilian observers and airport control. Two aircraft scrambled; both experienced avionics failures when attempting to engage. Object reported to divide into smaller objects. One of the most technically documented cases of the CUFOS era; cited in DIA report as an outstanding case. CUFOS and Hynek both cited this case in their late-1970s work as among the most significant radar-visual cases in the international record. Chapter reference: Chapter Nine (contextual). Sources: DIA report, September 1976; Wikipedia, 1976 Tehran UFO incident.

Note on case selection: This chronology covers cases discussed in the main text. The Blue Book files contain 12,618 total cases of which 701 are classified Unknown. The CUFOS UFOCAT database contains over 100,000 global reports. This appendix does not attempt to be a comprehensive case catalog but rather a reference tool for the narrative cases discussed in this biography. For the complete Blue Book case files, see the National Archives, Record Group 341. For CUFOS case files, contact the J. Allen Hynek Center for UFO Studies, Chicago.

Appendix C CUFOS Methodological Guidelines

This appendix summarizes the investigative methodology developed by J. Allen Hynek and formalized through the Center for UFO Studies, drawing primarily on Allan Hendry's The UFO Handbook: A Guide to Investigating, Evaluating, and Reporting UFO Sightings (Doubleday, 1979) and on Hynek's own methodological writing in The UFO Experience (1972) and The Hynek UFO Report (1977). The Handbook, based on Hendry's analysis of 1,307 cases received through the CUFOS UFO hotline during 1976–1977, represents the most systematic single application of CUFOS methodology ever published. It is the primary documentary record of how Hynek's investigative principles worked in practice. This appendix does not reproduce the Handbook's contents but summarizes its methodological framework for the reader's reference.

I. Foundational Principles

The null hypothesis. All cases begin with the assumption of conventional explanation. The investigator's first obligation is to apply every available conventional explanation rigorously before entertaining the possibility of a genuine anomaly. Hynek's twenty-one years of Blue Book experience established that the majority of UFO reports — his own estimate, consistent with Hendry's 1979 finding of 90 to 95 percent — resolve as IFOs (Identified Flying Objects) upon careful investigation. This finding is not a concession to skeptics; it is the methodological baseline that gives the residuals their significance.

The residuals concept. Cases that survive rigorous application of conventional explanations are not dismissed as inexplicable or relegated to the noise. They are treated as a dataset requiring analysis. The CUFOS methodology held that a genuine residual — a case surviving all conventional explanations after careful investigation by a qualified investigator — constituted a data point of scientific significance regardless of its content. Vallée's computerized analysis of the Blue Book files, conducted in collaboration with Hynek in the 1960s, demonstrated that these residuals are not randomly distributed. They exhibit systematic characteristics that distinguish them from noise.

Epistemological honesty. The CUFOS methodology required investigators to report what the evidence showed rather than what institutional, cultural, or personal preference required. This applied in both directions: investigators were not to dismiss cases that survived conventional explanation, and were not to endorse conclusions that the evidence did not support. Hynek described this principle in its simplest form in the 1953 JOSA article: Ridicule is not part of the scientific method, and people should not be taught that it is.

II. Case Intake and Initial Screening

Source evaluation. Cases came to CUFOS through the UFO hotline (operated via local police departments), direct witness contact, law enforcement referrals, and media monitoring. All incoming cases were screened before investigation was assigned. The screening applied a preliminary assessment of report quality: Was the observation of sufficient duration for a meaningful description? Were the observational conditions documented? Was the witness contactable for follow-up? Cases failing this preliminary screen were logged but not investigated.

Investigator assignment. Cases were assigned to investigators by area of relevant expertise. Optical cases — nocturnal lights, daylight discs, radar-visual cases with visual component — went to investigators with astronomical backgrounds capable of cross-checking against known optical phenomena. Cases with electromagnetic effects went to investigators with relevant engineering or physics backgrounds. Physical trace cases required investigators familiar with soil chemistry, vegetation analysis, and chain-of-custody evidence handling. Cases with physiological effects on witnesses required coordination with medical professionals.

Priority classification. Cases were prioritized by strangeness and credibility. A high-strangeness case from a single witness of unknown background was a lower priority than a moderate-strangeness case from multiple independent witnesses of documented credibility. The Hendry Handbook's analysis found that multiple-witness cases were actually more likely to be IFOs than single-witness cases in some categories — because IFOs capable of convincing many observers simultaneously tended to be dramatic objects (re-entering satellites, meteor fireballs, advertising aircraft) that also generated the most reports. This counterintuitive finding refined the CUFOS prioritization protocols.

III. Witness Evaluation

Professional background. Witnesses with professional training in observation — military pilots, law enforcement officers, air traffic controllers, astronomers, meteorologists — were assigned a higher prior credibility weight. This reflected not an assumption that trained observers were incapable of error but a recognition that they faced institutional disincentives to report anomalous phenomena: professional embarrassment, career risk, and community norms that valued rational precision. A trained observer who filed a report despite these costs was more likely to be reporting genuine experience.

Descriptive specificity. Witnesses providing precise, measurable descriptions — angular size, estimated distance, color gradations, duration, specific sequence of observed events — produced data that could be cross-checked against known phenomena and against other reports. Vague descriptions reduced the analytical value of a case. Investigators were trained to draw out specific observational details through structured interviewing rather than relying on witnesses' spontaneous narratives, which tended toward dramatization.

Narrative consistency. Witnesses were interviewed multiple times where possible, with care given to the stability of their accounts across sessions. Elaboration and embellishment — the tendency to add detail and dramatic significance in subsequent retellings — was a standard indicator of confabulation. Witnesses whose accounts remained spare, specific, and consistent across multiple interviews were demonstrating a form of memorial reliability that investigators were trained to recognize and document.

Recording. Hynek's son Joel, who as a young ham-radio operator recorded many of his father's witness interviews, described Hynek's insistence on recording as a core methodological principle. Verbatim records preserved the witness's own words, hesitations, and spontaneous corrections — data that transcription systematically destroyed. CUFOS protocols required taped interviews wherever witnesses consented.

Psychological state at time of sighting. Investigators documented the witness's psychological and physiological state during the observation: fatigue, emotional stress, intoxication, known visual impairments, prior expectations about UFOs. The Hendry analysis found that prior belief in UFOs did not reliably predict report quality in either direction — believers and skeptics produced both good and poor quality reports. The determining factor was observational opportunity and reporting discipline, not prior attitude.

IV. Conventional Explanation Tools

Astronomical cross-check. The first analytical layer applied to all cases. Investigators cross-checked the time, location, and described appearance of reported objects against: planetary positions (Venus was the single most common IFO stimulus across all categories); bright stars near the horizon (atmospheric refraction produces dramatic color changes and apparent motion); meteor trajectories and fireball characteristics; satellite passes and re-entry events (available through NORAD tracking data); and Iridium satellite flares. Hynek's astrophysics background established this as CUFOS's most powerful single explanatory tool.

Atmospheric optics. Temperature inversions, ducting, mirages, and unusual refraction conditions were evaluated against reported meteorological data for the time and location. Ball lightning and plasma phenomena, while rare, were assessed against reported weather conditions. Investigators were trained in the specific conditions that produced luminous atmospheric effects and in the documented behavioral characteristics that distinguished these phenomena from reports that survived atmospheric explanation.

Conventional aircraft. The Hendry analysis found that advertising planes — aircraft towing illuminated signs — were among the most significant IFO generators in the CUFOS dataset, producing reports that resisted explanation until the investigator established that an advertising flight had been operating in the area at the reported time. Military aircraft, experimental aircraft, and unusual flight formations were cross-checked through FAA records and military flight logs where available.

Psychological and perceptual factors. Autokinesis — the apparent motion of stationary lights in a dark visual field — was documented as a significant source of reported object maneuvers, particularly for nocturnal light cases. The Hendry analysis was notable for its iconoclastic finding that animal reactions, often cited as corroborating UFO reality, were equally produced by IFOs and therefore not diagnostic. Investigators documented observational conditions that could produce perceptual artifacts before attributing apparent behavior to the object itself.

V. Physical Evidence Protocols

Chain of custody. Physical trace evidence — ground impressions, scorched or dehydrated vegetation, soil samples, metallic fragments — required documented chain of custody from discovery through analysis. This meant photographing evidence in situ before collection, using standardized sampling containers, documenting the identity of all individuals who handled samples, and ensuring that laboratory analysis was conducted by credentialed facilities using documented procedures. The Val Johnson patrol car, maintained intact and transferred to the Marshall County Historical Society, represents the ideal case of physical evidence preservation.

Comparative sampling. Soil and vegetation samples from reported trace sites were compared against control samples from adjacent areas unaffected by the reported event. Statistically significant differences in soil chemistry, vegetation condition, or physical compaction between affected and control samples constituted independent evidence of a physical event at the reported location regardless of what caused it. The CUFOS protocols required control samples as a standard element of any physical trace investigation.

Electromagnetic effects documentation. Cases involving vehicle engine failure, compass deviation, radio interference, or electrical system anomalies required: documentation of the vehicle's prior mechanical state and maintenance history; comparison of the reported effect against known failure modes for that vehicle type; cross-checking of the timing, distance, and duration of the reported effect against the described object's behavior; and where possible, instrumental testing of the affected vehicle for residual anomalies. The Val Johnson case's Honeywell engineer and Ford Motor

Company examiner consultations are the exemplary application of this protocol.

Photographic evidence. CUFOS protocols required original negatives for any photographic assessment. Polaroid originals were required for Polaroid cases. Assessment without original materials was explicitly prohibited following the lessons of the Jaroslaw case (1967), in which Hynek's hedged public assessment based on copies produced misleading headlines that the originals, obtained nine years later, would have refuted. Photographic analysis involved: measurement of apparent angular size against known reference objects in the frame; optical geometry analysis to assess whether the image was consistent with the described object distance and size; and examination of the negative for evidence of multiple exposures or physical manipulation.

VI. Radar Case Protocols

Radar-visual cases — cases in which an unidentified object is simultaneously confirmed by independent radar returns and visual observation — were treated by CUFOS as the highest-priority category within the Close Encounter of the First Kind classification. The methodological reasoning was explicit: a visual witness may misidentify a conventional object; radar instrumentation cannot misidentify in the same way at the same time across multiple independent stations. Correlated radar-visual cases therefore eliminated the most significant category of conventional explanation — witness perceptual error — and placed the burden of explanation on the physical characteristics of whatever produced the radar return.

Radar case protocols required: documentation of the specific radar system type and its known performance characteristics; cross-checking with weather radar data for atmospheric ducting conditions that could produce false returns; confirmation of independent returns from multiple radar stations where possible; and comparison of the radar track against the visual description for consistency of position, heading, speed, and maneuver sequence. Cases with both radar and visual confirmation by trained military personnel — pilots, air traffic controllers — were considered the strongest available evidence category in the CUFOS case hierarchy.

VII. Hypnotic Regression — Special Protocols and Limitations

CUFOS treated hypnotically recovered testimony from abduction cases as data of a categorically different evidential status from consciously recalled field reports. Hynek documented this distinction explicitly in The Hynek UFO Report (1977): hypnotic regression produces material from the subconscious, which is not subject to the same consistency and specificity tests applicable to waking memory. The material may represent genuine experience, confabulation, or artifacts of the hypnotic process itself. The Lawson experiments, cited in Hendry's Handbook, demonstrated that hypnotized subjects who had not experienced any UFO encounter could produce detailed, consistent abduction narratives under hypnotic suggestion — a finding that severely limited the evidentiary value of regression-produced accounts.

Where hypnotic regression was conducted — as in the Pascagoula case with James Harder — CUFOS protocols required that regressed testimony be clearly distinguished from waking testimony in all case documentation and that published assessments specify which elements of the account derived from each source. The physiological and behavioral evidence independent of regression — the sheriff's secret recording, the polygraph results, the medical examination at Keesler AFB — carried the evidentiary weight in the Pascagoula assessment. The regression material was noted but not weighted equivalently.

VIII. Reporting Standards

CUFOS case reports distinguished between: (1) the witness's reported experience as documented; (2) the investigator's conventional explanation analysis and its outcome; (3) the evidence classification under the Hynek taxonomy; (4) the case disposition — explained, unidentified, or insufficient data. Reports were required to flag explicitly the evidential basis for each element of the analysis rather than presenting conclusions without supporting detail. The Val Johnson IUR coverage, which devoted nearly the entire September–October 1979 issue to a single case, represents the reporting standard Hynek considered appropriate for a high-quality unresolved case.

Public statements. Investigators were expected to limit public statements to conclusions supported by the available evidence. The standard for a public statement was higher than the standard for an internal case assessment — internal assessments could note the investigator's judgment about unexplained elements, while public statements required that judgment to be based on the physical evidence and witness testimony record rather than on the investigator's personal interpretation. Hynek's own Pascagoula public statement — these men have had a very real, frightening experience; they are absolutely honest — was deliberately limited to what the sheriff's recording, the polygraph, and the independent interviews actually demonstrated.

Primary sources for this appendix: Hendry, Allan. The UFO Handbook: A Guide to Investigating, Evaluating, and Reporting UFO Sightings. Garden City, New York: Doubleday, 1979. 297 pages. ISBN 9780385143486. Hynek, J. Allen. The UFO Experience: A Scientific Inquiry. Chicago: Henry Regnery, 1972. Chapters 1–4 (methodology and classification). Hynek, J. Allen. The Hynek UFO Report. New York: Dell, 1977. Chapter 1 (Blue Book methodology retrospective). CUFOS. cufos.org. Classification systems and investigation protocols.

Appendix D Hynek Classification Charts

J. Allen Hynek first presented his six-category classification system at the 134th Meeting of the American Association for the Advancement of Science, Boston, December 27, 1969, in his paper Twenty-One Years of UFO Reports. He elaborated the system fully in The UFO Experience: A Scientific Inquiry (Henry Regnery, 1972), chapter 10. The system divides UFO observations into two broad groups: distant sightings, in which the object is observed at range and the primary evidence is optical or radar; and close encounters, in which the observation occurs within approximately 500 feet of the witness and the range of available evidence expands accordingly. Within each group, categories are arranged by the type and quality of evidence available rather than by any assumption about the phenomenon's nature or origin. The system remains the standard taxonomy for UAP reporting in military, governmental, and scientific contexts worldwide.

Part One — Distant Sightings

Nocturnal Light (NL)

A light source observed in the night sky that cannot be attributed to any known natural or man-made phenomenon after investigation. No proximity threshold; the object is observed at distance. The diagnostic criterion is behavior inconsistent with known aircraft, satellites, planets, meteors, or atmospheric optical phenomena after systematic cross-checking against all available conventional explanations. Nocturnal lights are the largest single category in the Blue Book files and the most susceptible to conventional explanation — Venus alone accounts for more reports in this category than any other single source. Cases that survive full cross-checking despite the investigator's best effort to resolve them constitute the NL residuals.

Daylight Disc (DD)

An object observed in the daytime sky with defined shape and structure that cannot be attributed to any known natural or man-made phenomenon after investigation. No proximity threshold; the object is observed at distance. The diagnostic criterion is a defined form — disc, oval, cigar, sphere, or other — distinguishable from conventional aircraft, balloons, birds, or atmospheric phenomena after systematic cross-checking. The daylight condition, which eliminates many atmospheric optical explanations applicable to nocturnal light cases, makes the daylight disc category particularly resistant to conventional explanation when it survives analysis.

Radar-Visual (RV)

An object simultaneously confirmed by independent radar return and visual observation by at least one witness. No proximity threshold; dual confirmation is the defining condition. Atmospheric ducting and all known aircraft must be excluded. Multiple independent radar stations producing correlated returns constitute the strongest confirmation available in this category. Hynek considered radar-visual cases the highest-priority distant sighting category because radar instrumentation cannot misidentify in the same way at the same time across multiple independent stations — correlated radar-visual cases therefore eliminate the most significant category of conventional explanation, witness perceptual error, and place the burden of explanation on the physical characteristics of whatever produced the return.

Part Two — Close Encounters

Close Encounter of the First Kind (CE1)

A visual sighting of an unidentified object within approximately 500 feet (150 meters) of the witness, at sufficient proximity to preclude most conventional misidentification. The proximity threshold is the category's defining feature: at this range, the object's apparent angular size, structure, and behavior can be documented with sufficient specificity to permit rigorous cross-checking, and the range of conventional explanations that could account for the observation is substantially narrowed. Astronomical explanations, atmospheric optical phenomena, and most conventional aircraft are effectively excluded at 500 feet. The CE1 is the baseline close encounter — the threshold at which the investigator moves from evaluating a distant report to evaluating a proximate observation with correspondingly higher evidential expectations.

Close Encounter of the Second Kind (CE2)

A close encounter accompanied by measurable physical effects on the environment, on equipment, or on the witness's person, independent of the witness's subjective account. The CE2 is the category that most clearly distinguishes CUFOS methodology from purely testimonial investigation: physical evidence exists independently of what the witness says and can be examined, sampled, measured, and cross-checked by investigators who were not present at the event. Physical effects may include ground impressions or dehydrated vegetation at a reported landing site; interference with vehicle engines, electrical systems, or compass instruments; physiological effects on the witness such as temporary blindness, skin burns, or hair loss, verified by medical examination; and chemically or structurally altered soil in affected areas versus adjacent control samples. Hynek considered CE2 cases the most scientifically significant category because the physical evidence constituted independent confirmation not dependent on witness reliability. The Val Johnson incident of August 27, 1979 is the paradigmatic CE2 case in the CUFOS archive.

Close Encounter of the Third Kind (CE3)

A close encounter in which animate entities — beings of any described form — are observed in association with an unidentified object. Hynek deliberately chose the term animate entities rather than aliens, extraterrestrials, or occupants to preserve methodological neutrality with respect to the ultimate explanation. The category classifies the observation, not its cause: CE3 describes what the witness reported seeing, without any commitment to what those beings were or where they came from. Hynek treated CE3 cases as the highest-strangeness category and the one most requiring corroborating physical evidence before any published assessment. He withheld detailed discussion of CE3 cases from his 1969 AAAS presentation — introducing the taxonomy but deferring the CE3 elaboration to The UFO Experience — on the grounds that a skeptical scientific audience who had followed the argument this far on the strength of radar-visual and physical trace evidence cases would not follow it into territory involving humanoid occupants in the same session. The Pascagoula incident of October 11, 1973 is the best-documented CE3 case in the CUFOS archive.

Part Three — Additions After Hynek's Death

The following categories — CE4 and CE5 — are widely encountered in the UFO and UAP literature but were not part of Hynek's original six-category system. They are documented here to allow readers to distinguish Hynek's own taxonomy from subsequent additions attributed to it.

CE4 — Abduction (post-Hynek addition)

A case in which the witness reportedly is transported inside or aboard the object, against their will or without full conscious recollection. Associated primarily with Budd Hopkins and David Jacobs, who developed the abduction research framework in the late 1970s and 1980s. Hynek was aware of and interested in abduction reports but treated them as a methodologically distinct category requiring separate protocols because the primary evidence was typically obtained through hypnotic regression rather than waking recall. He did not incorporate abduction as a formal category in his published taxonomy. CE4 is not part of the original Hynek system.

CE5 — Bilateral Contact (post-Hynek addition)

A case involving voluntary, conscious, direct communication between a human witness and the object's presumed occupants, typically initiated by the human party. Associated with Steven Greer and the Center for the Study of Extraterrestrial Intelligence (CSETI). CE5 has no methodological basis in Hynek's published or archival work and is not part of the original Hynek system.

Part Four — The Two-Axis Assessment Framework

Beyond the six-category taxonomy, Hynek used a two-axis assessment framework for individual cases, described in The UFO Experience and applied throughout the CUFOS investigation protocols. The two axes were Strangeness and Probability.

Strangeness

The degree to which the reported characteristics of the observation depart from any conventional explanation. A nocturnal light behaving like a slowly moving aircraft is low-strangeness even if unidentified. An object that stops, reverses, performs right-angle turns, and produces electromagnetic effects is high-strangeness. Hynek argued that high-strangeness cases were paradoxically more scientifically significant, not less, because their resistance to conventional explanation was a property of the evidence rather than a failure of investigation. The Socorro landing was the paradigmatic high-strangeness case: physical trace evidence, specific and consistent witness description, no conventional explanation available after thorough investigation.

Probability

The assessed likelihood that the report represents genuine anomalous experience rather than misidentification, fabrication, or perceptual error. Probability drew on witness evaluation (professional background, descriptive specificity, narrative consistency across multiple interviews), corroboration (independent witnesses, physical evidence, radar confirmation), and the investigator's overall assessment of report quality. High probability combined with high strangeness defined the cases Hynek considered most scientifically significant: cases where the evidence for a genuine anomaly was strongest and the nature of that anomaly most resistant to conventional explanation. These cases received priority for field investigation and for publication in the CUFOS International UFO Reporter.

Sources

Hynek, J. Allen. The UFO Experience: A Scientific Inquiry. Chicago: Henry Regnery, 1972. Chapter 10 contains the definitive exposition of the six-category system with rationale for each category definition.

Hynek, J. Allen. Twenty-One Years of UFO Reports. In UFOs: A Scientific Debate, edited by Carl Sagan and Thornton Page. Ithaca: Cornell University Press, 1972. The original 1969 AAAS presentation introducing the taxonomy.

Hendry, Allan. The UFO Handbook: A Guide to Investigating, Evaluating, and Reporting UFO Sightings. Garden City, New York: Doubleday, 1979. The fullest practical application of the taxonomy and two-axis framework based on 1,307 CUFOS cases.

CUFOS. Classification Systems. cufos.org/types-of-ufos/classification-systems/. Authoritative current source for the taxonomy as maintained by CUFOS from 1973 to the present.

Made in the USA
Monee, IL
05 November 2026

67914368R00125